建 筑 绘 图

建 筑 绘 图

[美] 莱昂·克里尔 著
林 源 译

中国建筑工业出版社

著作权合同登记图字：01—2010—7052号

图书在版编目（CIP）数据

建筑绘图／（美）克里尔著；林源译.—北京：中国建筑工业出版社，2012.12
ISBN 978-7-112-14767-0

Ⅰ.①建… Ⅱ.①克…②林… Ⅲ.①建筑制图—图解 Ⅳ.①TU204-64

中国版本图书馆CIP数据核字（2012）第255963号

Copyright © 2009 Massachusetts Institute of Technology
All rights reserved.
Translation © 2012 China Architecture & Building Press
本书由MIT出版社授权我社翻译、出版、发行本书中文版
DRAWING FOR ARCHITECTURE / LÉON KRIER

责任编辑：戚琳琳
责任设计：陈　旭
责任校对：陈晶晶　赵　颖

建筑绘图
[美]莱昂·克里尔　著
林源　译
＊
中国建筑工业出版社出版、发行（北京西郊百万庄）
各地新华书店、建筑书店经销
北京嘉泰利德公司制版
北京云浩印刷有限责任公司印刷
＊
开本：850×1168毫米　1/32　印张：7$\frac{5}{8}$　字数：200千字
2013年6月第一版　2013年6月第一次印刷
定价：46.00元
ISBN 978-7-112-14767-0
（22798）

版权所有　翻印必究
如有印装质量问题，可寄本社退换
（邮政编码　100037）

目 录

序
作者札记

建造 ·· 1
城市/反城市：都市 VS 郊区 ················ 21
构图：有机的 VS 机械的 ····················· 67
两个世界 ······································· 77
建筑和都市的空间 ···························· 129
可持续的和不可持续的建筑高度 ·········· 149
转换的建筑与都市肌理 ····················· 163
形式与同一 ··································· 175
保存与维护 ··································· 191
建筑病理 ······································ 209

序

  2005年，我和莱昂·克里尔聊起关于我们如何生活的大问题正在开始给我们施加压力的话题。我所说的我们如何生活不仅仅是指当21世纪蹒跚而来的时候人类物质性的居住地将会怎样，还有这种文明的存在本身是否会继续——我所说的文明是针对我们的城镇而言的。

  在那个时候，一种认识在西方兴起，那就是现代文明的基础将很快受到持久的全球性能源危机的威胁。用一个术语来简要概括这个危机就是"石油峰值"，它提供了若干直截了当的解释和更多的暗示。很显然，据说燃烧矿物燃料的工业时代的接近尾声是由于完全的能源耗尽（并且这个关于所谓的不同寻常的能源图景的消息所引起的疑虑是其他任何可能会弥补那些损失的事情无法消除的）。在此之外，"石油峰值"暗示了许多我们的现代生活赖以存在的复合系统要尽早建立，一旦这个世界的石油生产达到峰值我们就要开始减少使用——似乎正像是现在这样的状况。"复合系统"是从我们制造食物的方式到我们进行贸易的方式，到我们居住在土地上的方式、我们活动在土地上的方式。在金融方面，资金的配置和管理是一个尤其敏感的复合系统，即使在我写这些的时候它也正在全球性地发生崩裂。这在很大程度上是由于石油峰值预示了常规化的周期性工业增长的结束和纸质证券交易上的有效表现。

  石油峰值的状况对于克里尔来说是意义重大的，因为这是对他发起的在建筑学与城市规划领域重建传统实践的运动的令人信服的全新支持——也就是说，广泛的现代主义规划的持续性将很快在实践上成为不可能。赋予20世纪特征的广义的都市肿大已经是由石油和天然气来负担了。从摩天楼到美国的幸福郊区都会在石油供给崩溃之后失去价值和作用。为建筑的现代主义者所喜爱的资源和能源密集型的建造材料——金属板材、梁、桁架、高技术的饰面、钢筋混凝土和环氧胶——会变得越来越稀缺。通过装配式方法只能在构造上表达为古典主义的某些种类，返回到地区性的自然材料才会是必然。新开发的增长会必然回复到一个同人的精神和生理更为适合的规模。城市这个系统的规模必须要减小。石油峰值为克里尔的观点提供了真正的迫切性。他的现代主义的反对者们的意识形态不论是以形而上还是以高级时尚为框架的，在这些现实的辩论面前都萎缩了。

回溯至20世纪70年代，从克里尔职业生涯的早期开始，克里尔的立场一般都同新的资源紧张保持着密切的关联。25年里他一直用流利的、强有力的论辩来痛骂现代主义的过度，在勒·柯布西耶和密斯·凡·德·罗的追随者们面前呈现了新鲜的、得体的、多样的景象。他在类型学的训练方面是特别全面和清晰的——类型的语汇已经被现代主义抛弃到了历史的垃圾船上，已经沦为可悲的语意不清了，因为没有了建筑和城市的融合。

克里尔的论辩著作总是伴随着草图、图解和涂鸦，带着绝妙的精炼和睿智来表明他的观点。他藏在符号和"艺术的创造力"的沙袋后面，这对于试图穿透反对者的精神防线的战士来说是一个好策略。在这些图中，逐字的解说被直接的更高级的大脑认知活动所代替，精神上的侵权行为将被更有说服力的图像证据所消解。克里尔的草图在同现代主义的蒙昧所导致的混乱作斗争方面特别的有效。他的草图的威力源自于我们承认这个持久的现实依然存在，并且延续着，不用管那些宣称他们创造了他们自己的现实的人的狡黠的诡辩。

克里尔的草图经常采用辩证的形式，呈现两个相对立的观点之间的联系：工业的VS手工业的，古典的VS乡土的，现代的VS传统的，等等。其他的例子使用的是图像速记法，这种手段相较于事实陈述的影响更大。他用图像讨论标准的柏林城市街区的缺点及其补救办法，把一个复杂的事件有效地用一页的篇幅来说明。

在克里尔的草图和文字中，这些对抗和矛盾都是作为自我认识的论辩。克里尔总是在进行某种辩论并持有明确的立场。这些立场来自于一种内在的、综合的世界观，一种对非同寻常的技术必胜主义的渐弱回归的理解，这种技术必胜主义定义了20世纪及其知识运动，增加了审美的困扰，接着就是由此而引发的对我们人类居住场所的破坏。克里尔的工作实际上是对原则的一系列的澄清和纠正——因为这些基本的东西一直在被建筑的学术官僚和城市规划的行政管理所混淆、滥用和错用，在我们的时代，去创造值得我们付出感情的建筑和场所已经变得不可能了。当然，所有这些混乱和滥用的结果在美国是遍地可见的，并且在其他国家也是越来越多，那些条状林荫带、停车用的荒地、摩天楼、分割的土地、单一模式的"住房"，以及这个正在结束的时代里其他令人感到凄凉的东西。

莱昂·克里尔的工作已经对所有这些做出了极好的矫正，虽然他有时是孤身战斗，立足于现代主义的十足的惯性使他在改革的尝试遇到抵抗时变得更强大。在这本新书中，草图坚定地说明了它们自己，无需他总是强有力的散文的协助。它们极端的清晰和简明能对抗任何误解。它们越过抽象捍卫事实，越过虚假捍卫真相，越过仿造捍卫有机，越过乏味捍卫优雅。克里尔在一系列的连续画面中展示他的观点，每一个都是富于想象力的。

因为它们都是图画，也许会给年轻一代留下图像胜于文字的雄辩这一印象。这些年轻的从业者所继承的世界将远不是舒适、便利的，这是第二次世界大战结束后西方工业时代的标准。我认为莱昂·克里尔和我都很平静地确信这一点，那就是我们都处在大的变化中。我预感到这些新的状况将迫使我们对日常生活的所有活动都作出另外的安排。它们将改变我们栖息于大地的景观，它们将会要求我们去重新建立起与被遗弃的建筑和城市的传统法则的联结。莱昂·克里尔的新书将成为关于几乎已经到来的未来的一系列卷帙中的一部。

詹姆士·霍华德·昆斯勒
Saratoga Springs，纽约

作者札记

在我的建筑与城市设计之外，我还制造了大量的涂鸦和符号，这就是本书的主题。它们的产生不是自然而然的，只是突然的、短暂的，还常常是一种愤怒的爆发。它们经常汇集为一或两个图像，是我以前在设计方案、写作或演讲中想要清晰表达的东西。这在1980年第一次发生，在一个延续很久的艰苦写作之后，我在几天之内打了两百多页字的同时也积累了几乎一样多的这种涂鸦。其中大部分都有克制的潜在意义；实际上，这是与持久的侵犯、荒谬以及矛盾的对抗，生活也因此而丰富无比。

这些图都是未经加工的、不拐弯抹角的，不是安慰或取悦的手段，而是要揭示建筑的实践和思想体系中令人愤慨的东西，它们概括地使用概念化的工具来重新建立传统的城市与建筑。而这些行动所引发的反应，如大家所知，大部分都是出乎意料的或毫无意识的。在人类活动终结时什么将会胜利呢，是愿望得以实现的生活还是令人沮丧的生活？是好的还是不好的，是对的还是错的，是明智的还是愚蠢的？显然，不论是人类智慧还是人工智慧都没有获得对于终极问题的内在认识。然而，我们不能再忽视环境、建筑物和能源政策、工业文明正处在悲剧性的困境里这一重要问题了。我们在错误的地方建造，以错误的方式、错误的材料、错误的类型、错误的密度和高度建造，而且居住者的数量也是个错误。

在我看来，在化石燃料时代的传统建筑及其建造与环境技术是全球性的生态重建的有效工具。就像在过去，这是大自然的状态，将对我们的发展可能性重新进行定义。而这些涂鸦，也许有助于指明我们思考的方向。

莱昂·克里尔

建造

vernacular Building 乡土的 建筑物

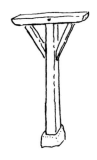  

CLASSICAL ARCHITECTURE 古典的 建筑

尺寸・类型・表达
SIZE × TYPE × EXPRESSION

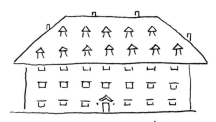

Cottage PALACE-SIZE
"COTTAGE LOOK"
农舍 "宫殿的尺寸"
"看上去是农舍"

PALACE Cottage-size
"PALACE LOOK"
宫殿 "农舍的尺寸"
"看上去是宫殿"

VERNACULAR × *classical*
乡土的 × 古典的

USE & MISUSE
使用 & 非使用

PALACE ~ PALACE-SIZE
宫殿~宫殿的尺寸

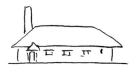

Cottage ~ Cottage size
农舍~农舍的尺寸

CLASSICAL & *vernacular*
古典的 & 乡土的

LK 03

建造

 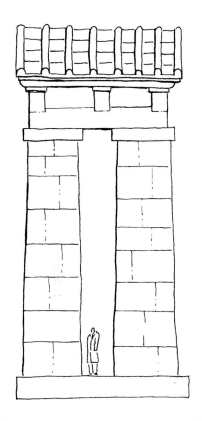

MIDGET "monument"	极小的 "纪念性建筑"	MONSTER "SHED"	巨大的 "棚屋"
vernacular scale MONUMENTAL RETHORIC	普通建筑的规模 纪念性的修辞	monumental scale VERNACULAR PROSE	纪念性建筑的规模 普通的散文
EFFETE PRETENTIOUS INAPPROPRIATE	衰弱的 自命不凡的 不适合的	GROSS POMPOUS WEAK	茁壮的 虚夸不实的 虚弱的
MINIATURE Architecture	小型的 建筑	MONUMENTALIST BUILDING	纪念性的 建造物

HUMAN DIMENSIONS	人的尺度	SUPERHUMAN DIMENSIONS	超人的尺度
DOMESTIC SCALE	自身的规模	MONUMENTAL SCALE	纪念性建筑的规模
TECTONIC LOGIC	技术的逻辑	ARCHITECTONIC LOGIC	建筑技术的逻辑
ARTISANAL "MATERIALOGICAL"	匠人的 "唯物论的"	ARTISTIC "MATERIALOGICAL"	艺术家的 "唯物论的"
TECHNOLOGICAL MIMESIS	技术的模仿	SYMBOLIC ENCODED MIMESIS	象征的编码的模仿
RES PRIVATA	个人的东西	RES PUBLICA	公共的东西
Building	建造物	ARCHITECTURE	建筑

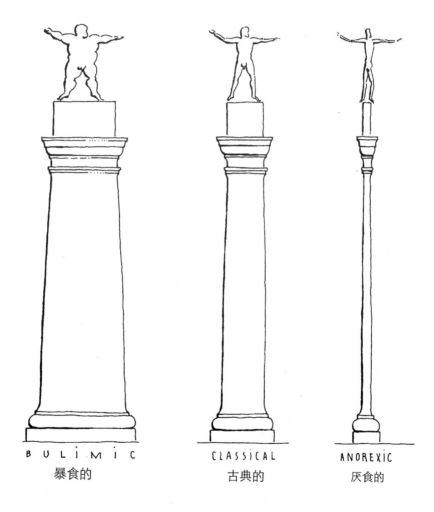

古典的　　　　比例　　　变形

CLASSICAL PROPORTIONS DISTORTED

A 古典的塔斯干柱式
A CLASSICAL TUSCAN ORDER

竖向比例随意变化
VERTICAL DIMENSIONS VARIED ARBITRARILY

B LAPIDOCEPHALITE
B 柱身变化
C CLUB·FOOTED
C 柱脚变化

DIAMETERS VARIED ARBITRARILY
直径随意变化

D 厌食
D ANOREXIC
E BULIMIC
E 暴食

建造　7

有什么事情出错了
something went wrong
ON THE BUILDING SITE
在建造现场

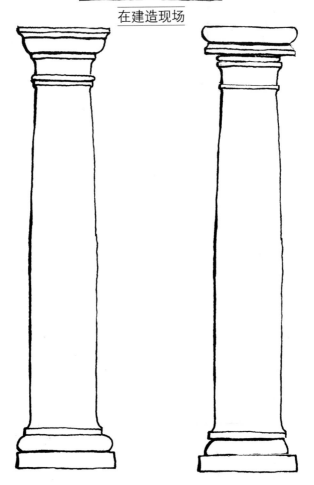

"Not to worry ~ Nobody will see the difference"
"没有错～没有人会看到两者的差异"

CLASSICAL
古典的

ACLASSICAL
非古典的

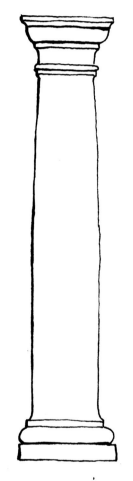
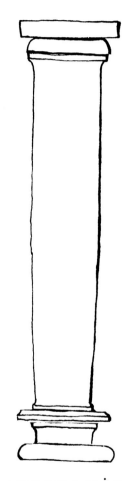

TECTONIC
PARTS MARRIED
构造的
各部分的混合

ATECTONIC
PARTS DIVORCED
非构造的
各部分的离散

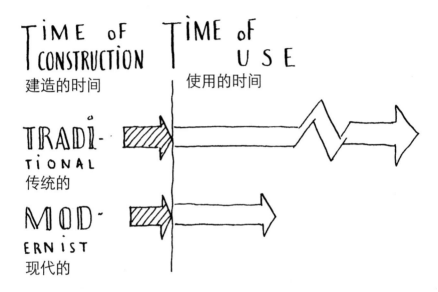

THE SOCIAL EFFICIENCY OF A BUILDING-METHOD IS MEASURED BY RELATING A BUILDING'S TIME OF USE TO ITS TIME OF CONSTRUCTION

建造方式的社会效能的计量是与使用的时间和建造的时间相关联的

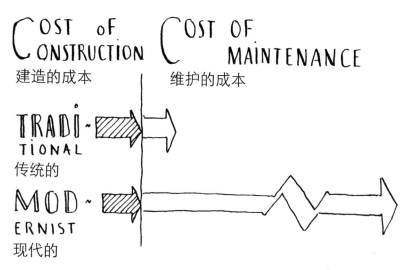

THE ECOLOGICAL and ECONOMIC COST-EFFICIENCY of A BUILDING METHOD is MEASURED by DIVIDING THE ADDED COST of CONSTRUCTION and LONG-TERM MAINTENANCE By THE NUMBER of USE-YEARS

建造方式的生态与经济的成本效能的计量是要区分建造的附加成本和使用年限的长期维护成本的

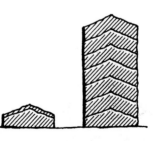

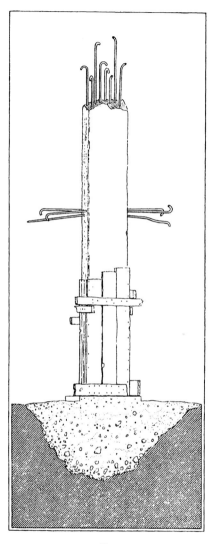

THE 6TH ORDER
OR
THE END OF ARCHITECTURE
L.K. 77

第六种柱式
或
建筑的终结

结构的一元 与 形式的多元
结构文化的统一

CONSTRUCTIONAL MONISM & STYLISTIC PLURALISM
UNITY OF CONSTRUCTION CULTURE

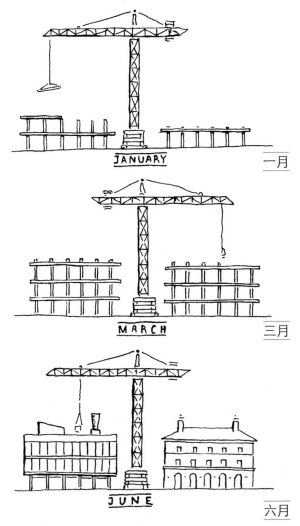

LAST MINUTE SURPRISE ON BUILDING-SITE
LK 84
建造现场的最后一分钟惊喜

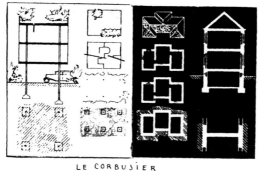

LE CORBUSIER
5 POINTS OF A NEW ARCHITECTURE
AGAINST TRADITIONAL BUILDING

勒·柯布西耶
新建筑五点
不同于传统建筑的

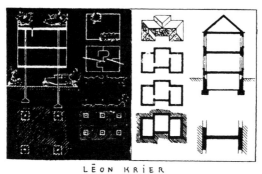

LÉON KRIER
5 POINTS OF TRADITIONAL BUILDING

莱昂·克里尔
传统建筑五点

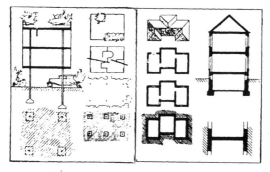

10 POINTS OF PRESENT BUILDING
现在的建筑十点

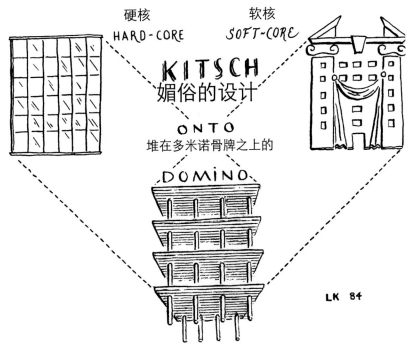

DISTANCE & DETAIL
距离 & 细节

建造

I am a house
我是一个房子

Call me a house
叫我房子

I am a window
我是一个窗子

Call me a window
叫我窗子

I am a house-door
我是一个大门

Call me a house-door
叫我大门

I am a roof
我是一个屋顶

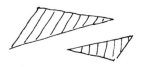

Call me a roof
叫我屋顶

LK 82

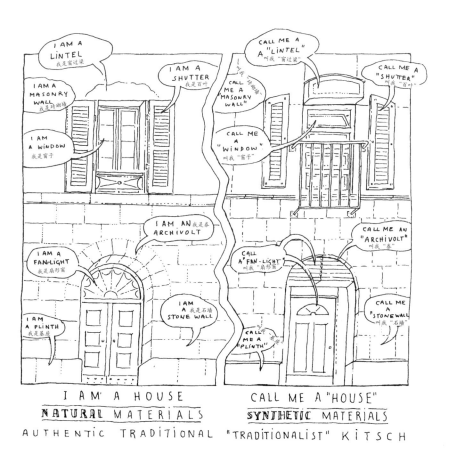

我是一个房子　　　　叫我"房子"
天然的材料　　　　　合成的材料
真正的传统　　　　　"传统主义"的媚俗设计

体块与空间的比例和谐
PROPORTIONAL HARMONISATION of VOLUMES and VOIDS

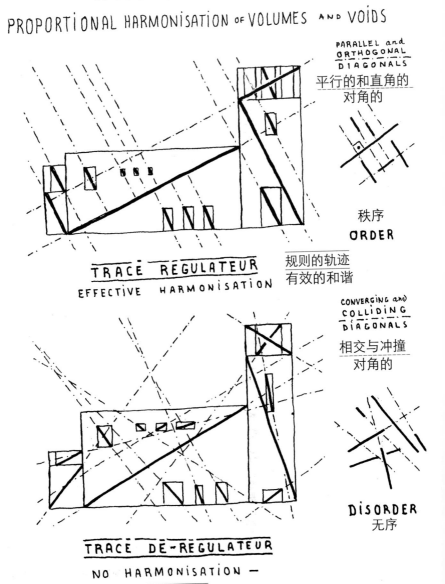

城市/反城市：都市 VS 郊区

功能区域的
反城市

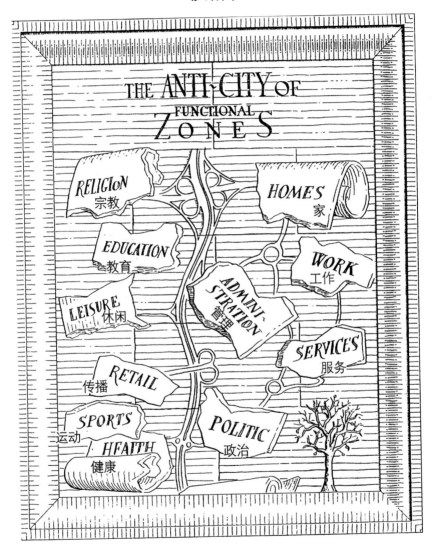

都市社区的城市

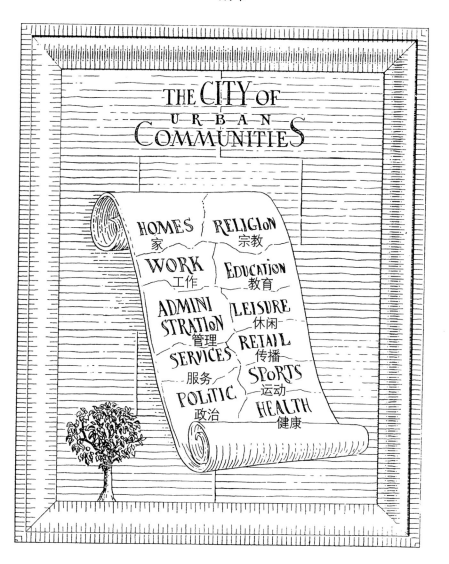

城市/反城市：都市 VS 郊区

功能区域的反城市

都市社区的城市

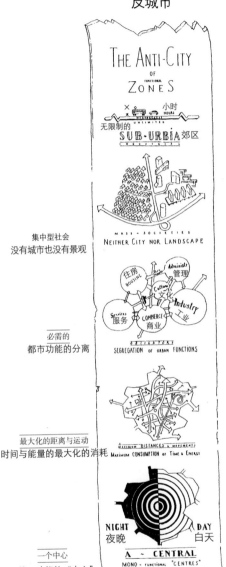

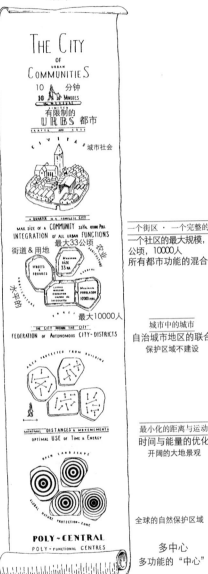

集中型社会
没有城市也没有景观

一个街区 · 一个完整的城
一个社区的最大规模，33
公顷，10000人
所有都市功能的混合

必需的
都市功能的分离

城市中的城市
自治城市地区的联合
保护区域不建设

最大化的距离与运动
时间与能量的最大化的消耗

最小化的距离与运动
时间与能量的优化
开阔的大地景观

全球的自然保护区域

一个中心
单一功能的"中心"

多中心
多功能的"中心"

城市
CITY

VARIABLE NUMBER
OF
COMPLETE
URBAN COMMUNITIES

完全都市社区的可变数量

反城市
ANTI-CITY

VARIABLE NUMBER
OF
MONO-FUNCTIONAL
ZONES

单一功能区的可变数量

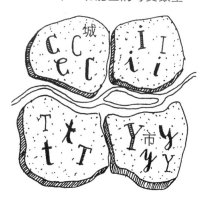

城市/反城市：都市 VS 郊区

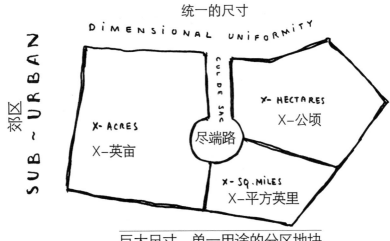

MEGA-SIZED SINGLE-USE ZONED LOTS
巨大尺寸 单一用途的分区地块

SOCIAL, FUNCTIONAL AND ARCHITECTURAL **MONOTONY** 社会的、功能的和建筑单调

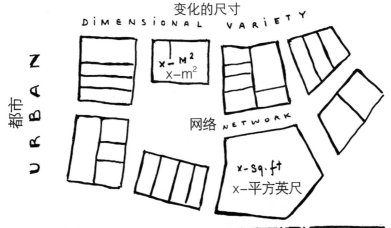

VARIOUSLY SIZED + USED BUILDING LOTS
多样的尺寸 + 建筑用地

SOCIAL, FUNCTIONAL + ARCHITECTURAL **VARIETY**

社会的、功能的和建筑的多样性

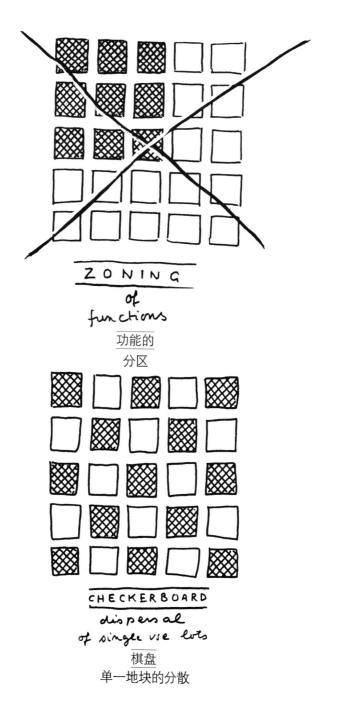

城市/反城市：都市 VS 郊区

SETBACKS ENFORCE BUILDING AWAY FROM PLOT-BOUNDARY 强制性的后退远离用地的边界建造	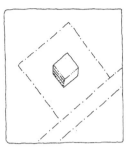 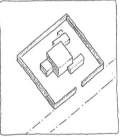 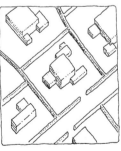	ALIGNEMENTS ENFORCE BUILDING ON The BOUNDARY OF THE PLOT 强制性的定位在用地的边界建造
ONLY REALIZING A WEAK DIFFERENCI- ATION OF PRIVATE AND PUBLIC REALM 仅仅实现了私密与公共领域的微弱的区分		FRONTAGES ESTABLISH A PHYSICAL DISTINCTION BETWEEN PUBLIC AND PRIVATE REALM 临街的建造私密与公共领域之间的实体性的区
URBAN GROWTH INCREASES THE FEELING OF DISORDER OF THE ENVIRONMENT 都市的生长加强了环境的无序感		URBAN GROWTH INCREASES THE FEELING OF ORDER AND URBANITY 都市的生长加强了秩序感和都市感

SUB~URBAN URBAN

郊区 都市

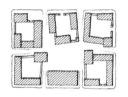

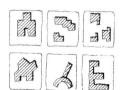

城市/反城市：都市 VS 郊区

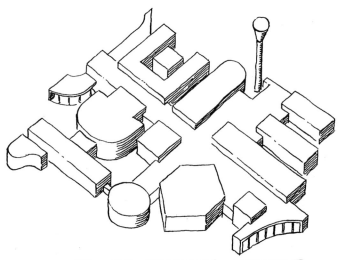

CONCENTRATION OF CIVIC USES
城市设施的集中

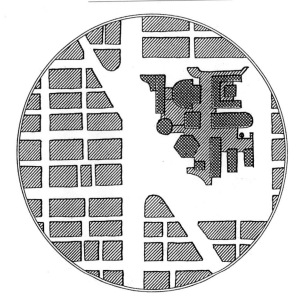

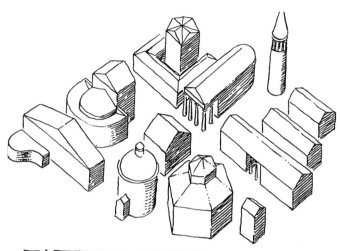

DISPERSAL OF CIVIC USES
城市设施的分散

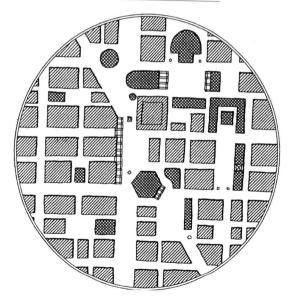

城市/反城市：都市 VS 郊区

身体的分区

功能的分离 ⟶ 感官世界的分解

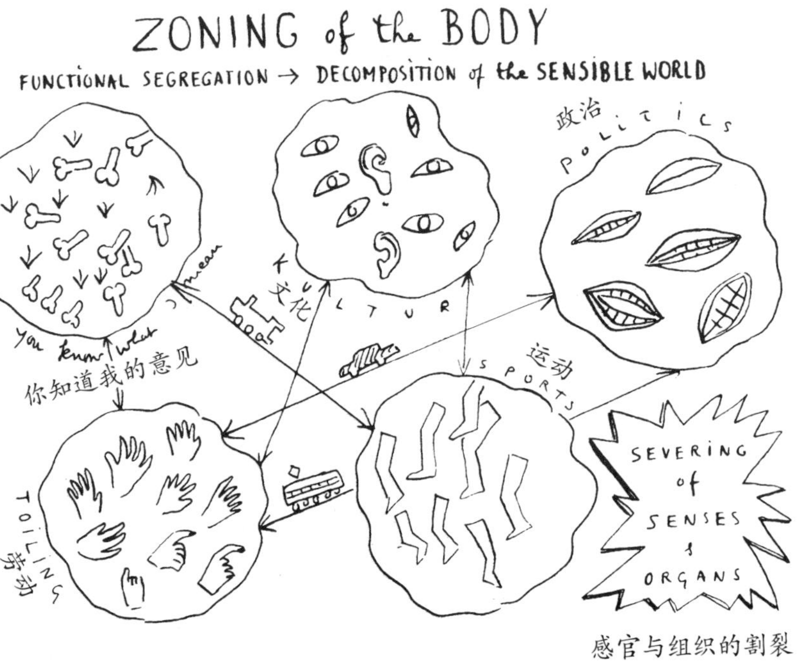

ZONING of the BODY
FUNCTIONAL SEGREGATION → DECOMPOSITION of the SENSIBLE WORLD

- 政治 POLITICS
- KULTUR 文化
- 运动 SPORTS
- TOILING 劳动
- you know what I mean 你知道我的意见

SEVERING of SENSES & ORGANS

感官与组织的割裂

节约的生活
在无序蔓延的城市中
化石燃料时代

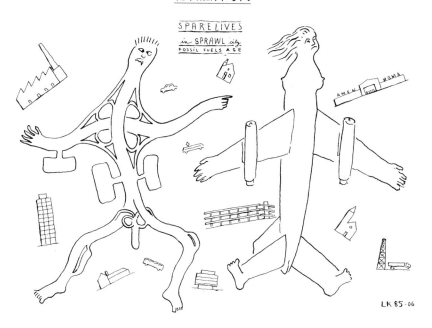

"分区"的观念
求助于每周一次的人类个体的进入
世界中心委员会1号指令

The Idea of "Zoning"

as applied to the weekly gastronomical intake of an individual of the human species

WORLD CENTRAL COMMITTEE DIRECTIVE N°1

monday	~	32 Pints	of	Liquids................
Tuesday	~	3 Kg	of	Meat...................
Wednesday	~	2,5 Kg	of	Fats....................
Thursday	~	3 Kg	of	Pasta...................
Friday	~	2 Kg	of	Fish....................
Saturday	~	6 Pints	of	alcoolic Drinks........
Sunday	~	6 ℔	of	Backery................

mond........ Note Individual deceased.... Experiment discontinued

星期一～ 32品脱液体…… ……

星期二～ 3公斤肉… …… ……

星期三～ 2.5公斤脂肪… …… ……

星期四～ 3公斤面条… …… ……

星期五～ 2公斤鱼… …… ……

星期六～ 6品脱酒精饮料… …… ……

星期… …注意 个体已经中止…体验不连续

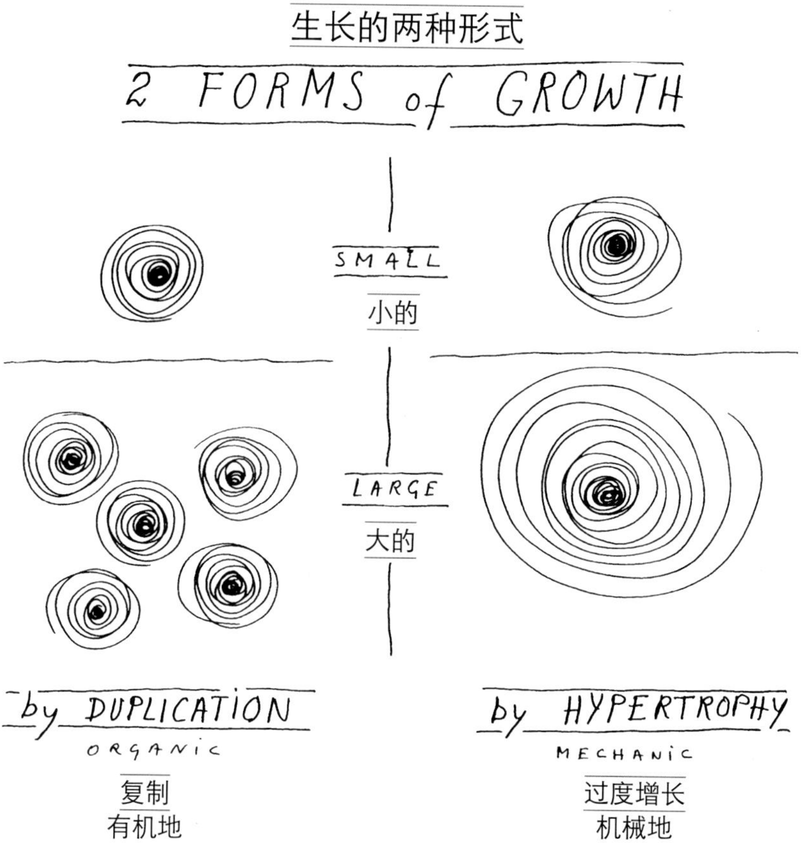

生长的两种形式

2 FORMS of GROWTH

SMALL
小的

LARGE
大的

by DUPLICATION
ORGANIC

复制
有机地

by HYPERTROPHY
MECHANIC

过度增长
机械地

一个功能区
使得城市具有单一的特性（功能），排斥其他所有的

A FUNCTIONAL ZONE admits one single quality (function) of a City at the exclusion of all others

EXCLUSIVE
排他

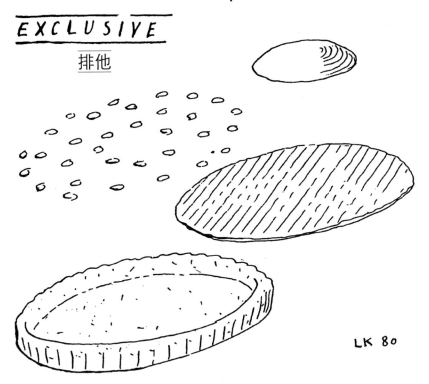

LK 80

All that is not specifically obligatory is strictly forbidden
任何不具有明确的必要性的东西都要严格禁止

城市街区
包含着并激发了一个城市的所有特性

An URBAN QUARTER CONtains and PROmotes all the Qualities of a CITY

IN-CLUSIVE

包含

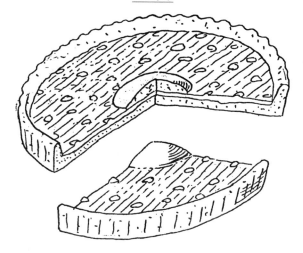

All is Permitted & Promoted that is not strictly forbidden

所有的东西都是被允许的、被推动的
没有严格的禁止

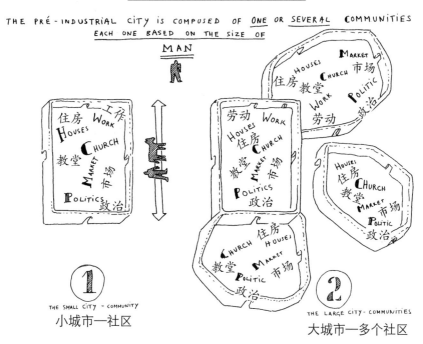

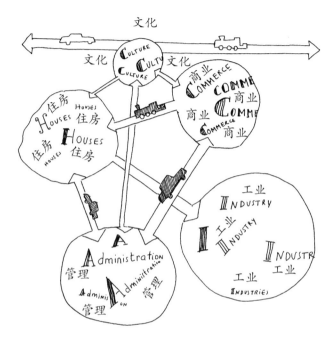

THE INDUSTRIAL CITY IS DECOMPOSED INTO ZONES

工业城市分解成区域

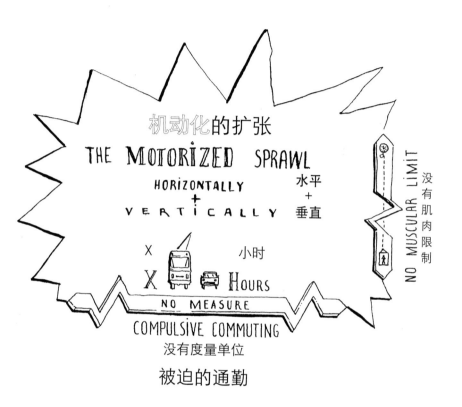

成功都市的风险

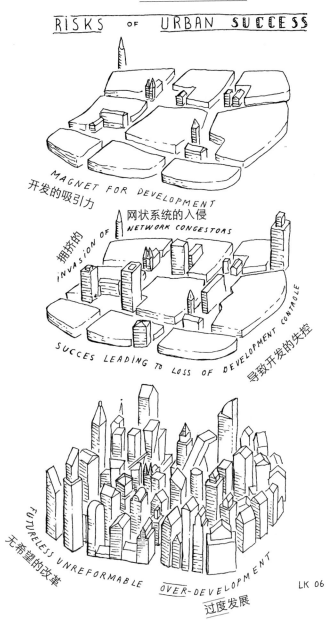

竖直 & 水平
都市生长 和 拥挤
生长·过度生长·高速生长 等等

VERTICAL & HORIZONTAL
URBAN GROWTHS and CONGESTORS
GROWTH · HYPERGROWTH · WEEDS ETC

MATURE
HORIZONTAL NETWORK
WITH VERTICAL PEDESTRIAN
TWIGS (SHORT CUL-DE-SACS)

成熟
带竖向步行道路分支的水平网状结构
（短的尽端式道路）

TOWN
城镇

MATURE
HORIZONTAL NETWORK
USURPED BY OVERSIZED MECHANIZED
CUL-DE-SACS (WEEDS)
NETWORK CONGESTORS

成熟
水平网状结构被过度发展的
尽端式机动交通所侵占（蔓生的）
网络的拥堵

DOWN-TOWN
市中心

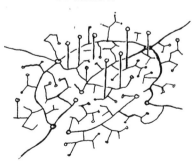

SPRAWLING LABYRINTH
MECHANIZED VERTICAL AND
HORIZONTAL CUL-DE-SACS · MAXIMUM
GEOGRAPHIC CONGESTION

蔓延的迷宫
机动的竖向与水平的尽端式道路
地理的拥堵

SPRAWL
蔓延

反都市的迷宫
ANTIURBAN LABYRINTH

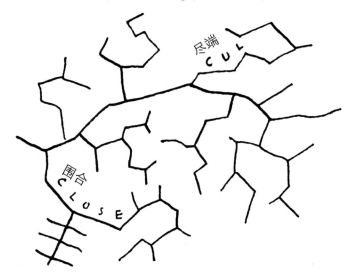

SUB~URB
城郊地区

FAU~BOURG
近郊（靠近城市出口的高速
道路的近郊聚居地）

BAN~LIEU
城市外围

VOR~ORT
有自己的商业中心的郊区

TRABANTEN~STADT
卫星城

TOWNSHIP
乡镇

SATELLITE
卫星镇

THE ARSE OF THE WORLD
世界的末端

都市模式
URBAN PATTERN

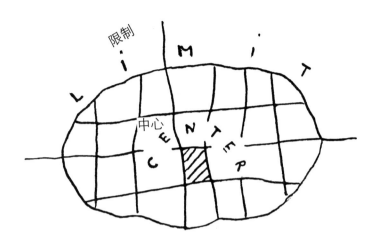

URBS
都市
BOURG
社区
LIEU
独立区域
ORT
地方
STADT
城市
TOWN
镇子
CITY
城市
CENTER OF THE WORLD
世界的中心

都市社区的城市

标准度量单位
步行
多中心的融合

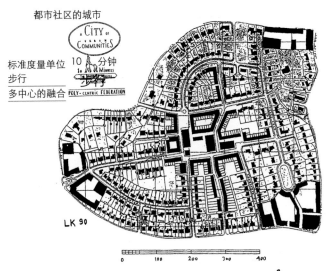

REDEVELOPMENT through FUNCTIONAL and TYPOLOGICAL MIX

重建 通过功能与类型的混合

功能区域的反城市

强迫性的通勤
单中心的卫星城

TYPICAL SEGMENT OF 100% RESIDENTIAL SUBURB
⟵ 最大33公顷，步行10分钟 ⟶
100%的郊区住区的典型片段

大型用地的正确的区划
CORRECT ZONING OF LARGE EMPLOYMENT

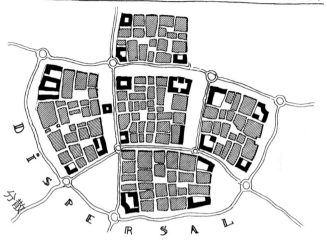

DISPERSAL 分散

MISTAKEN ZONING OF LARGE EMPLOYMENT
大型用地的错误的区划

市中心　BUSINESS-PARK 商业园区
DOWN-TOWN
SUB-URBIA　郊区
CONCENTRATION 聚集

城市/反城市：都市 VS 郊区

The CITY as COMMUNITY(IES)

CULTURAL, POLITICAL, ECONOMICAL
RELIGIOUS ⇨ RELIGATA

作为社区的城市
文化，政治，经济
宗教的 ⇨ 约束

The CITY
COMMUNITY
CIVITAS
PARISH
etc

城市
社区
全市
教区
等

The BIG CITY
大的城市．
FEDERATIO
of COMMUNITIES etc
社区等的联合

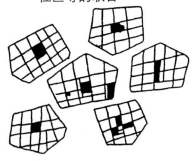

MONO-CENTRIC
单一中心

POLY-CENTRIC
多中心

10 MINUTES
10分钟

N × 10 MINUTES
N个10分钟．

The CITY'S LIMIT is a BUILT one
城市的界限是一个建造的界限

The INDUSTRIAL Anti-CITY
A-CENTRIC
ANTI-URBAN — ANTI-RURAL

工业的反城市
中心
反都市的-反乡村的

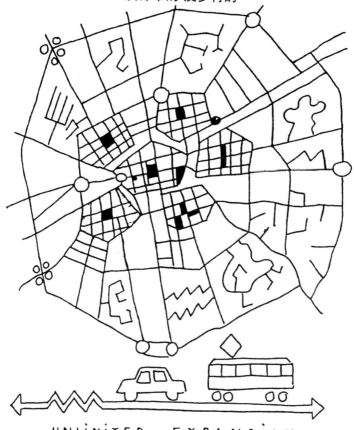

UNLIMITED EXPANSION
无限地扩展

The CITY'S LIMIT is a BUILT on
城市的界限是一条行政管理的边线

城市/反城市：都市 VS 郊区

步行的城市
The CITY of the PEDESTRIAN

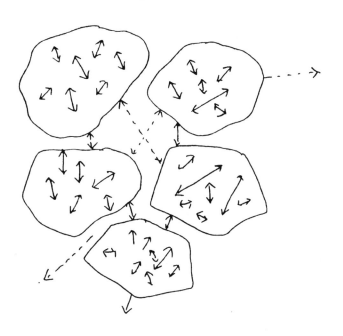

MINIMUM DISTANCES
FOR MAXIMUM ACHIEVEMENT & PLEASURE
最小的距离
为了保证最大程度的抵达与愉悦

机动交通的（反）城市
The (Anti-) CITY of the MOTORCAR

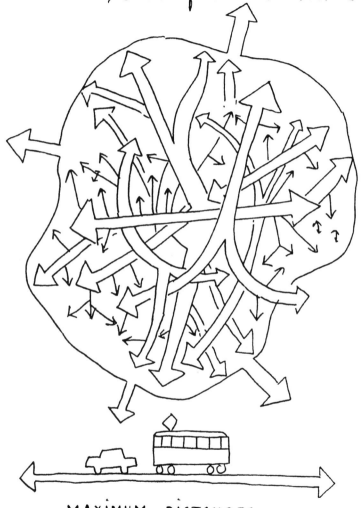

MAXIMUM DISTANCES AND BOREDOM for MINIMUM ACHIEVEMENT

最大的距离
最小程度的抵达造成的乏味

correct density and composition
= nameable CITY

混合的密度与构图
＝名副其实的**城市**

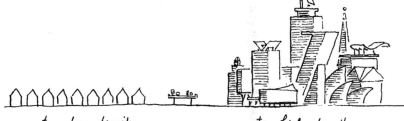

too low density　　　　too high density

太低的密度 ＝ socalled "CITY" 太高的密度
　　　　＝所谓的"城市"

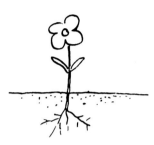

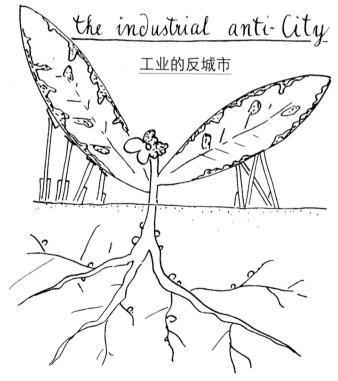

OVER-development of CENTER
UNJUSTIFIABLE PROFIT-MAKING
中心的过度发展
不合理的收益——自己造成的

Urbanization of SUB-urb !
JUSTIFIED PROFIT MAKING
郊区的都市化！
形成合理的收益

FUNDAMENTAL CHOICES OF URBAN DEVELOPMENT
都市发展的基本选择

the city
城市

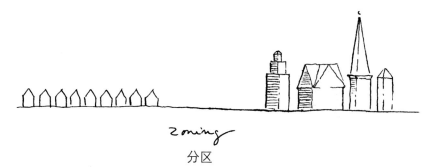

zoning
分区

城市/反城市：都市 VS 郊区

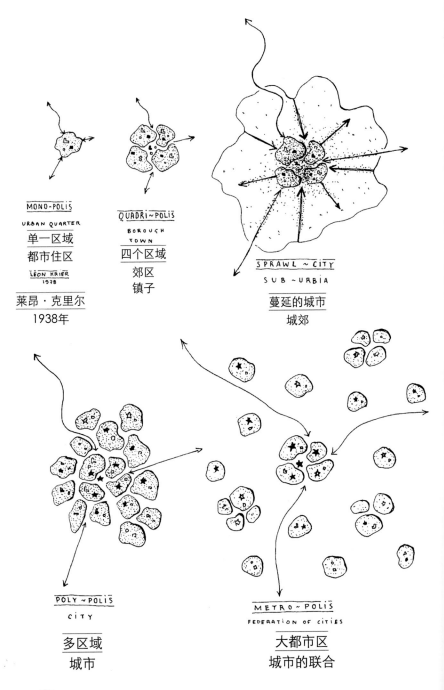

城市 & 附着物
CITY & PARASITE

CITY without SUBURB
城市 没有郊区

CITY with SUBURB
城市 有郊区

SUBURB without CITY
郊区 没有城市

CITIES within the CITY
城市群 在城市中

郊区 不是都市。 城市外围 不是中心。 近郊不是城里。 卫星城不是城

SUBURB no URBS ∘ BANLIEU no LIEU ∘ FAUBOURG no BOURG ∘ VORORT no ORT

THE HANGMAN and its VICTIM
THE ANTI-CITY is out to KILL the CITY

字母游戏 和 它的受害者
反-城市 就是要杀了城市

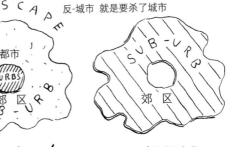

CITY
and
LANDSCAPE
a good
MARIAGE

this CiTY is no longer
a true CITY
this Landscape no longer
a true LANDSCAPE
SUBURB ALWAYS
DEFEATS BOTH

a SUBURB
WITHOUT a
CITY
WILL FIND ITS VICTIM
at whatever EXPENSE
EFFORT or DISTANCE

城市　　　这个城市不再是一个真正的城市　　　郊区

和　　　　　　　　　　　　　　　　　　　　没有

景观　　　这个景观不再是一个真正的景观　　　城市

　　　　　　　　　　　　　　　　　　将找到它的受害者

良好的　　　　　　　　　　　　　　　无论花费多少的

结合　　　郊区总是让它们都落空　　　努力 或 距离

1850~1950年
工业的反城市的形成
（工业城市＝目的的相互抵触）

1850 ~ 1950
THE FORMATION of the INDUSTRIAL ANTI-CITY
(INDUSTRIAL CITY = CONTRADICTIO in TERMINI)

INDUSTRIAL
SUB URB
BAN LIEU · FAU BOURG
VOR ORT ·
SATELLITES · TRABANTEN
BESIEGE
the CITY

工业的
郊区
城市外围 · 近郊
卫星城 · 卫星镇
包围
城市

The CITY is
FINE
WITHOUT SUBURB

SUB~URB
UNTHINKABLE

WITHOUT the CITY

没有郊区的城市是
美好的

———

没有城市的郊区是
不可思议的

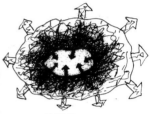

SUBURBS
FIRST
DESTROY the
LANDSCAPE & FORESTS
AND THEN
the
CITY

郊区
首先
毁掉了
景观和森林
然后就是
城市

城市/反城市：都市 VS 郊区

机械化的～交通
功能分区的效果

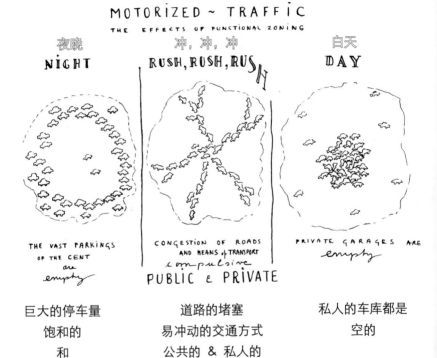

巨大的停车量
　饱和的
　　和
　　空地

道路的堵塞
易冲动的交通方式
公共的 & 私人的

私人的车库都是
　空的

人与化学能量的浪费

反城市的性别
⇨ 工业化的效果
生产地点，行政管理～消费的社会性的分离

THE SEXES in the Anti-CITY
⇨ THE EFFECTS OF INDUSTRIALISATION
SOCIAL SEGREGATION of PLACES of PRODUCTION, ADMINISTRATION - CONSUMPTION

NIGHT 夜晚　　　　　　　DAY 白天

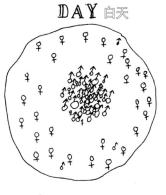

DAILY
FRUSTRATION
SOCIAL ISOLATION
FUTILE
CONSUMPTIONS
SOCIAL
DISINTEGRATION

ALIENATION IN MINI-FAMILIES
DICTATORSHIP of MASS-MEDIAS

ALIENATION IN PLACES OF
PRODUCTION & ADMINISTRATION

小家庭的疏离
大众传媒的专权

日常的
沮丧
社会性的
孤独
没有意义的
消费
社会的
瓦解

生产与管理在地点上的疏远

单一中心的 · 大都市

多中心的 · 联合体

致命的武器不是汽车而是郊区的住宅

NOT THE CAR BUT THE SUBURBAN HOME IS THE DEADLY WEAPON

DAILY SUBURBAN MORTARFIRE AGAINST URBAN CENTERS

日常的郊区往返是对都市中心的严重打击

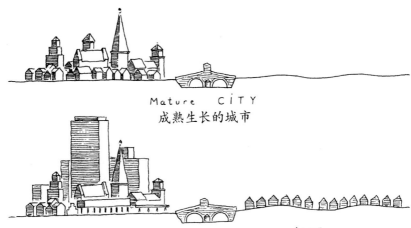

Mature CITY
成熟生长的城市

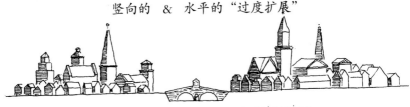

Vertical & Horizontal "OverEXPANSION"
竖向的 & 水平的"过度扩展"

Organic EXPANSION through DUPLICATION
通过复制的有机扩展

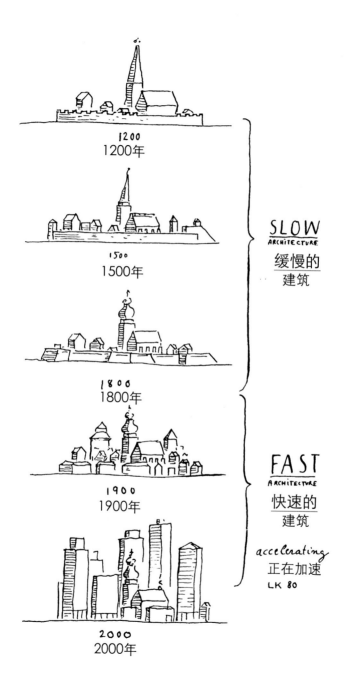

城市/反城市：都市 VS 郊区

构图：有机的 VS 机械的

构图的样式

类型学的　　　　　机械论的

Modes of Composition

TYPO~LOGICAL　　MECHANISTIC

POLYCENTRIC SYMMETRIES
SEPERATE
多中心的对称
分离

MERGED

POLY·AXIAL SYMMETRIES
TYPOLOGICAL ASYMMETRY

FORM
合并
多轴线的对称
类型上的不对称
形式

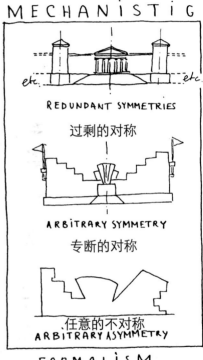

REDUNDANT SYMMETRIES
过剩的对称

ARBITRARY SYMMETRY
专断的对称

ARBITRARY ASYMMETRY
任意的不对称

FORMALISM
形式主义

The MATERIAL of the COMPOSITION
构图的原材料

The COMPLETE COMPOSITION
ORGANIC
完整的构图
有机的

The REDUNDANT SYMMETRIES
MECHANICAL   LK 84
过剩的对称
机械的

构图：有机的 VS 机械的

I love her so, oh mother!
我如此爱她，噢妈妈！

classical
古典的

Ilovehersoohmother!
我如此爱她噢妈妈！

vernacular
乡土的

love
IIII lovehersososo ohmother
ohmother

我我我如此如此如此爱爱爱她噢妈妈噢妈妈噢妈妈！

academical, mechanical, capricious
学院的、机械的、强词夺理的

I 我 e 爱 oooo hh 她 r ! eee 噢

industrial (zoning)
工业的（分区）

伪装的对称
False SYMMETRIES

构图：有机的 VS 机械的

正确的 和 伪装的 对称
CORRECT and FALSE SYMMETRIES

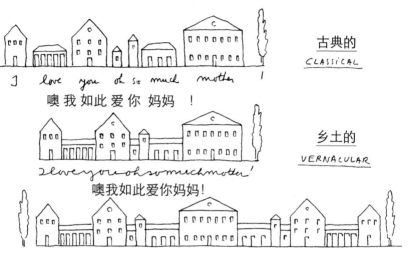

古典的 CLASSICAL
噢我如此爱你妈妈！

乡土的 VERNACULAR
噢我如此爱你妈妈！

机械的 mechanical
噢我如此爱你妈妈！爱你妈妈噢我如此爱你妈妈噢我如此

NATURAL SYMMETRY
自然的对称
NATURAL ARTISTIC ORDER
HIERARCHY
自然的艺术的秩序
层级结构

Ego

(UNIQUE) IDENTITY　（唯一的）特征
INDIVIDUAL　个体

Vernacular Crafts　NATURE　Classical Arts
Natural Hierarchy

地方工艺　自然　古典艺术

Part and Whole
部分与全体的自然的层级结构

MECHANICAL SYMMETRY
机械的对称
TYRANNICAL ORDERING
A-HIERARCHICAL
强制的排序
某种-层级的

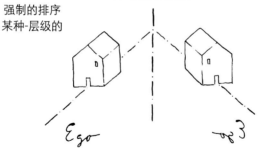

Ego　　　　　*Ego*

IDENTICAL IDENTITY　　同一的特征
CLONE　　克隆
Academic & Industrial　学院的 & 工业的
Tyrannical SUB-ordination
of
Part under Whole or Whole under Part
全体之下的部分的 或者 部分之下的全体的
强制的次级排序

构图：有机的 VS 机械的　73

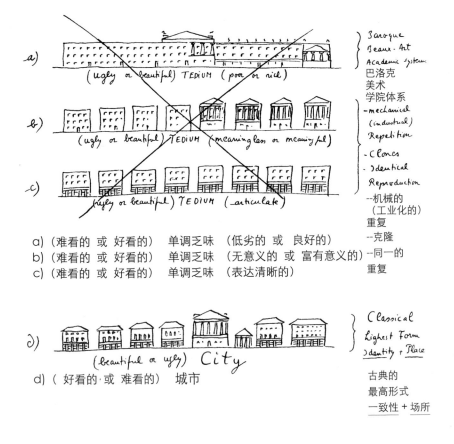

a) （难看的 或 好看的） 单调乏味 （低劣的 或 良好的）
b) （难看的 或 好看的） 单调乏味 （无意义的 或 富有意义的）
c) （难看的 或 好看的） 单调乏味 （表达清晰的）

d) （好看的 或 难看的） 城市

层次结构的复杂性 内容 = 形式	反层次结构的 复杂性	无层次结构的复杂性 内容 × 形式
HIERARCHICAL COMPLEXITY CONTENT = FORM	ANTI-HIERARCHICAL COMPLEXITY	NON-HIERARCHICAL COMPLEXITY CONTENT X FORM

ORGANIC ORDER & SYMMETRY CLASSICAL ORDER RATIONALISM AS MEANS INDIVIDUAL	MECHANICAL ORDER + SYMMETRY BAROQUE-ACADEMIC → ORDERING → RATIONALISM AS END CLONED →	PLANNED DIS-ORDER & COERCION MODERNISMS, PUNK, DECOMPOSTION as STYLE RATIONALISM as STYLE INDIVIDUALISTIC {HIGH BROW / COMMERCIAL / TECHNO} KITSCH
有机的秩序 & 系统 古典的 秩序 作为手段的理性主义 个体的	机械的秩序 + 系统 巴洛克～学院的 排序 作为目标的理性主义 克隆	计划的无秩序 & 强制 风格上的分解 风格上的理性主义 个体特征 {高额 / 商业化 / 技术} 俗气的艺术

构图：有机的 VS 机械的

两个世界

两个世界
The two Worlds

Crafts & Arts
ecological industries

技术 & 艺术
生态的产业

The World
as a solid, permanent practical and beautiful house (for) of mankind

这个世界
是一个可靠的，恒久的，实实在在的，美丽的家，（为了）属于人类的

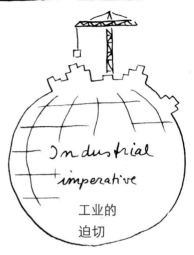

Industrial imperative

工业的
迫切

The Planet
an endless, trivial messy & noisy building-site

这个星球
一个没有尽头的、微不足道的、肮脏的、嘈杂的建造场地

人体工业
HOMO INDUSTRIALIS

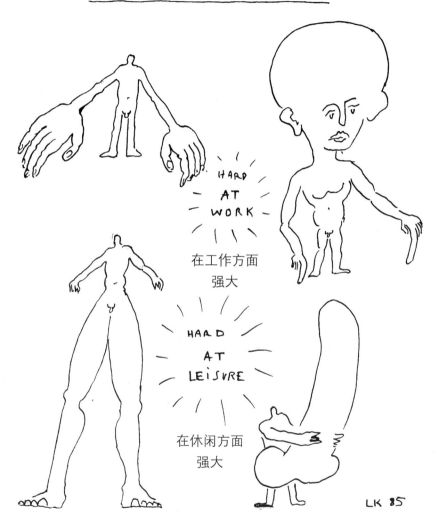

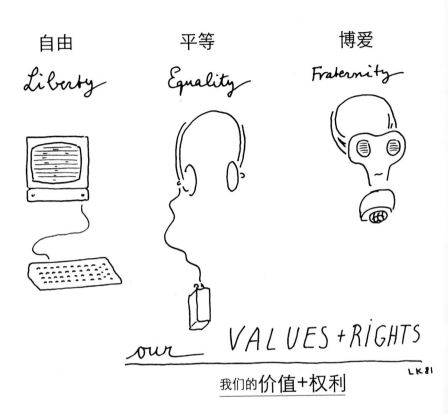

我们的工业
La vie industrielle

family stress

家庭的压力

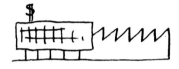

Work stress

工作的压力

age stress

年龄的压力

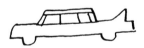

"I am at peace"

"我现在平静了"

两个世界

高技术 和 低技术
HIGH and LOW TECH

"THERE IS NO GOING BACK...dear"
(wishfull thinking)

"没有回去······亲爱的"
（痴心妄想）

"MODERN" and "DATED"
FORMS OF HANDTOOLS

"现代的" 和 "过时的"
手工的形式

工业 ~ 文化
INDUSTRY ~ CULTURE

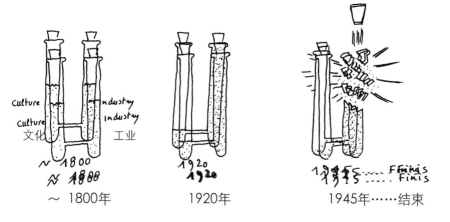

~ 1800年　　　　1920年　　　　1945年……结束

Culture from then on with a Capital C...

文化 从那时起是大写的……

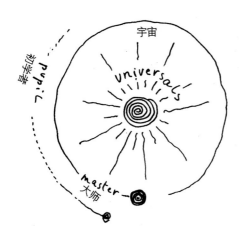
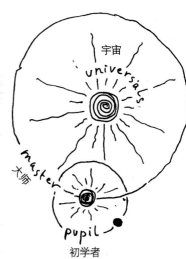

the good teacher
好的老师

the bad teacher
坏的老师

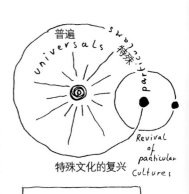

AUTHENTIC CULTURE
真正的文化

CULTURAL REVIVAL
文化的复兴

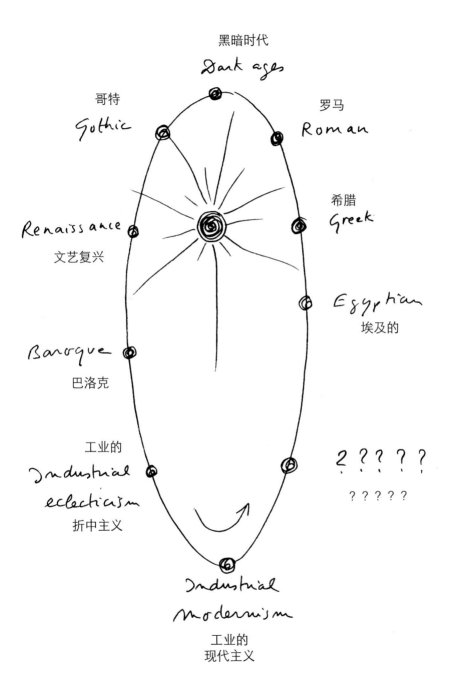

两个世界 85

l'architecte de l'ère machiniste
建筑师 VS 机械师

LEON KRIER · 16 BELSIZE PARK · LONDON N·W·3
莱昂·克里尔 · BELSIZE公园16号 · 伦敦

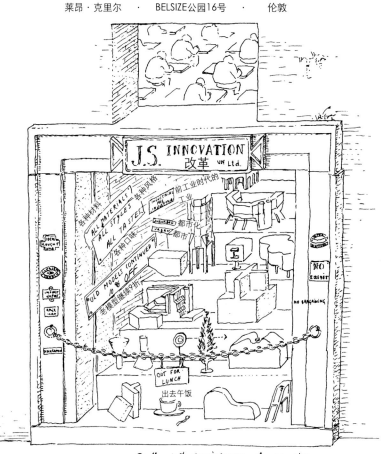

牢固关系的10周年纪念
7月28日.1968～1978年.莱昂

*
轻便的，坚定的，完美的

两个世界 87

STYLES OF CONTEMPORARY ARCHITECTURE COMMISSIONS in most "developed countries"

当代的风格
建筑设计任务
多数"发达国家"中的

PUBLIC BUILDINGS
公共建筑

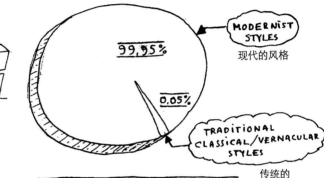

PRIVATE RESIDENTIAL
私人住宅

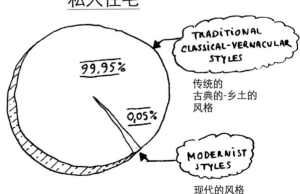

LK 88

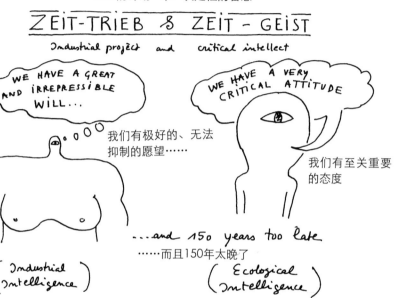

两个世界

工业的第二个最重要的成就　　　　　　　　　　集体主义

去讨论、分解、批评的愿望是与思考、构成、建议的无能成比例增长的

The 2nd Greatest Achievement of ~~Industry~~ Collectivism

THE WILL TO DISCUSS, DECOMPOSE, CRITICIZE GROWS IN PROPORTION TO INCAPACITY OF THINKING, COMPOSING, PROPOSING

我感到害怕　　　　　　打击胜过麻木　　　　　　我们是……

我想表达些东西，我不知道是什么　　　　　妈的　　　　　我是如此重要

让我们讨论吧　　批评　　提供的　　我们要参与　　去做什么？

MILLIONS of INTELLIGENT BRAINS PERMANENTLY INCAPACITATED TO THINK AND WORK INDEPENDENTLY of INDUSTRIAL IDEOLOGY

成千上万个智慧的大脑永远不具备思考和工作的能力
工业化思想意识的独立性

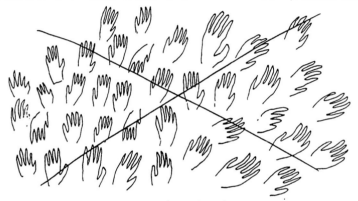

手工劳动的价格同需要赡养的失业的手的数量成正比

通过训练有素的集体无知和集体失业永久地排除和消解了的工业大都市及产品的潜在竞争

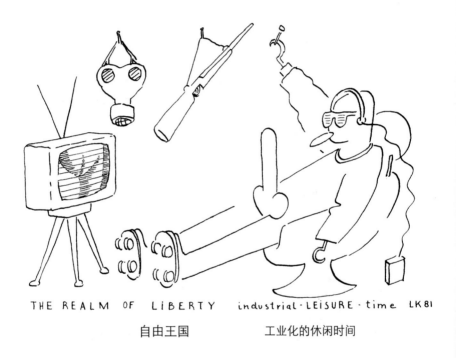

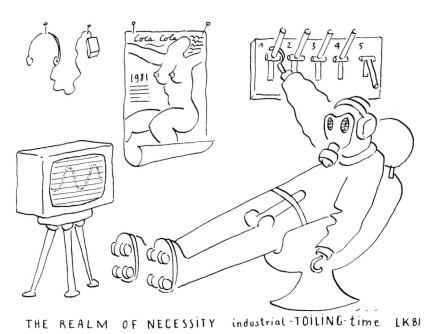

必然王国　　　　　　工业化的劳作时间

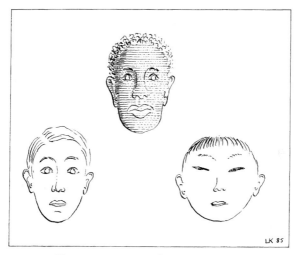

Traditional Pluralism
传统的多元主义

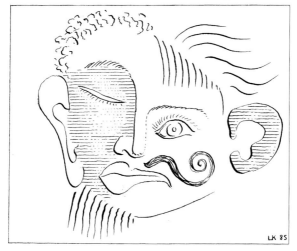

modernist Pluralism
现代派的多元主义

两个世界 95

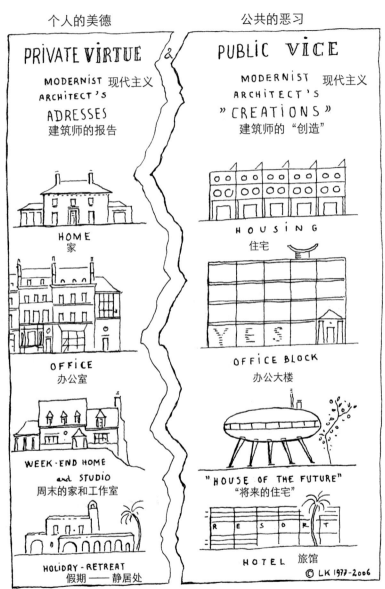

房地产代理处　　　　建筑学校

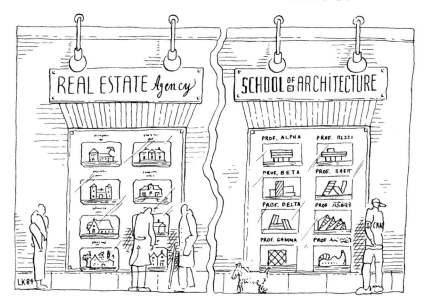

两个世界

"SPIRIT OF TIME" WINNING AGAINST "SPIRIT OF PLACE"
("Cutting-edge" spirit being a child of cosy environment)
"时间的精神"战胜了"场所的精神"
("锐利的"精神是温暖舒适的环境的产物)

MODERN Traditional
现代 传统的

MODERN Traditional & Modernist
现代　传统的&现代主义的

MODERN Modernist
现代　现代主义的

两个世界 101

BEAUBOURG ?
美吗?

BEAUBOURG !
美!

LK 93

ACTUAL + SYMBOLIC 真实的 + 象征的
VALUES 价值
natural - artificial 自然的 - 人工的

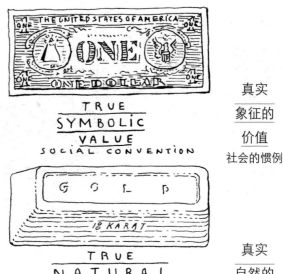

TRUE
SYMBOLIC
VALUE
SOCIAL CONVENTION

真实
象征的
价值
社会的惯例

TRUE
NATURAL
VALUE
CHEMICAL FACT

真实
自然的
价值
化学的事实

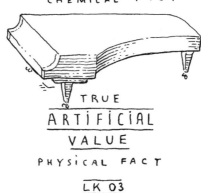

TRUE
ARTIFICIAL
VALUE
PHYSICAL FACT

真实
人工的
价值
物理的事实

LK 03

累积的两种形式

Two Forms of Accumulation

ARCHITECTURE (WITH SCULPTURE)   SCULPTURE   PAINTING
建筑(有雕塑的)   L.K. 88   雕塑   绘画

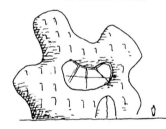

SO~CALLED ARCHITECTURE   SO~CALLED SCULPTURE   SO~CALLED PAINTING
所谓的建筑   所谓的雕塑   所谓的绘画

两个世界

没有出错……回到……童年
当我长大的时候我要变小就好像我是爸爸……

舒服的式样 和
良好的尺寸
COMFORT ~ MODEL and
good measure

传统 / traditions

按尺寸定做
智慧的惯例　made to measure
INTELLIGENT CONVENTION

现代主义 / modernismo

"创造性"
速写 ~ 偏执的惯例　"CREATIVE"
SCHIZZO ~ PARANOÏAC CONVENTION

LK 83

两个世界

MIES·CENTENNIAL·PROPOSAL
FUTURE OF MIES-ANTHROPY
MORE MIES IS LESS
MIES PRIVATE FURNISHINGS PROMINENTLY DISPLAYED IN "MIESLAND"-SHRINE

密斯·百年纪念·方案
密斯的 ～人类学的 未来
更多的密斯就是少
密斯的私人家具显著地展示在"密斯神坛"上

两个世界

你不能向后转这个轮子……
你不能向后……年轻人……

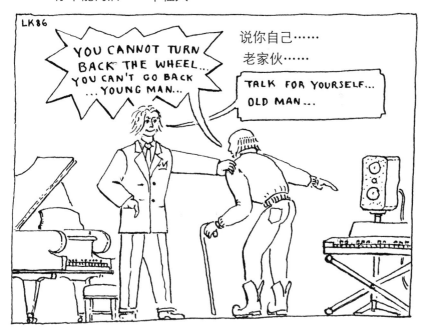

说你自己……
老家伙……

THE URBAN MAN
都市的人

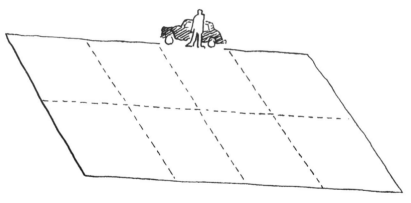

THE SUB-URBAN MAN
郊区的人

MAN · VEHICLE · LANDNEED
人 · 交通工具 · 土地需求

FROM SMALL TO LARGE FAMILY (I)
从小家庭到大家庭（I）

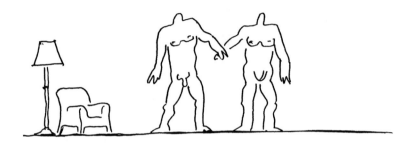

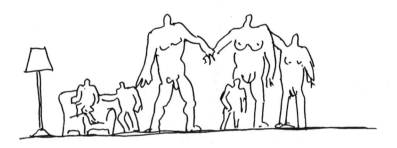

ORGANIC FORM OF GROWTH
MULTIPLICATION
生长的有机形式
增加

FROM SMALL TO LARGE FAMILY (II)
从小家庭到大家庭（II）

MONSTROUS FORM OF GROWTH
OBESITY
生长的庞大形式
肥胖

正确的 都市 生长
不成熟的人生长直到成熟

CORRECT URBAN GROWTH
IMMATURE QUARTERS GROW TILL MATURITY

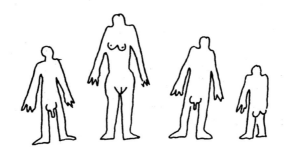

CATASTROPHIC ANTI-URBAN GROWTH
MATURE PARTS DEGRADE INTO FRAGMENTS

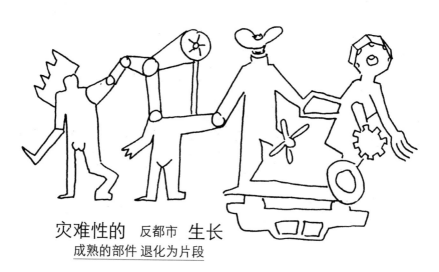

灾难性的 反都市 生长
成熟的部件 退化为片段

都市生长

有机的 VS 机械的

Urban GROWTH
ORGANIC versus MECHANIC

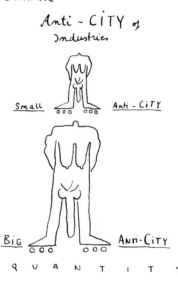

城市 工艺和艺术	父母 不要生长 但是要增加 （上帝说）	工业的反城市	
小城市 一个社区	增加 （不要生长） 社会大众 建筑师 生长 通过分区 在整个城市的消耗方面	小的	反城市
大城市 一个有多个社区的家庭		大的	反城市

质量 VS 数量

两个世界   119

哥伦布因素
The Columbus Factor

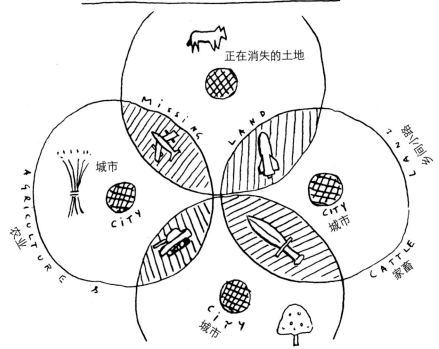

X Number of Cities need X times more <u>Land</u> For whatever <u>Land</u> they are missing they are going to bash in their own heads and rather than reduce their own numbers, they are going to invade, conquer and subjugate far <u>Lands</u>, continents and peoples

城市所需的<u>土地</u>数倍于城市的用地。对于任何<u>土地</u>来说它们都正在失去它们要打坏它们的头而不是要减少它们的数量，它们要侵入，征服，降伏更多的<u>土地</u>、大陆和人民。

City & Landscape

城市 & 景观

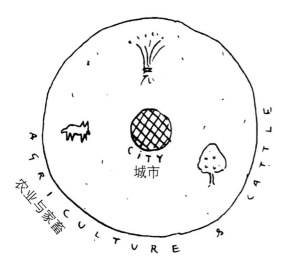

A CITY needs APPROX so much Land for its nutrition

一个城市大约需要如此之多的土地，以保证土地的营养供给。

LK 1982

两个世界

世界商业中心的50000个居民
燃料消耗的体积

50'000 OCCUPANTS OF WORLD TRADE CENTER
VOLUME OF FUEL CONSUMPTION
IN 31 MONTHS AT RATE OF 2006
(7 BILLION BARRELS PER YEAR FOR 230'000'000 CITIZENS)

31个月，2006年的水平
（23000万居民，每年70亿桶）

→ 1 MONTH CONSUMPTION OF 50000 OCCUPANTS
50000个居民一个月的消耗

24 个月 / 12 个月 / 31 个月
31 MONTHS / 24 MONTHS / 12 MONTHS

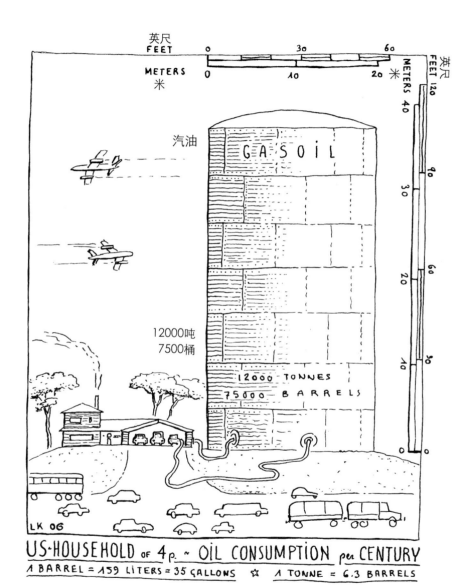

全球变暖

GLOBAL WARMING

CONSEQUENCE OR CAUSE

结果 或者 原因

WHAT CAUSES
什么引起的

BRAIN WARMING

大脑变暖

而且让我告诉美国居民
一件事……为了节约能源我们准备好去做任何事情；
但是我能让你们所有人都相信一件事，……我们
将永远不会放弃我们美国人的生活方式
A.Schlesinger
世界能源大会，华盛顿，1977年

大脑的力量　&
（局限）

BRAIN power &
(LIMITED)

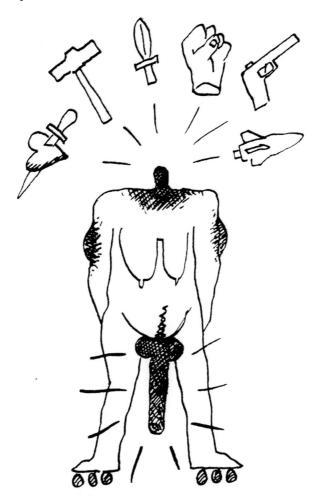

HOMO INDUSTRIALIS
人体工业

音乐的力量
(无限)

MUSCLE power
(UNLIMITED)

$$\begin{pmatrix} N \times 1\,Bp = 1\,Bp \\ N \times 1\,Mp = N\,Mp \end{pmatrix}$$

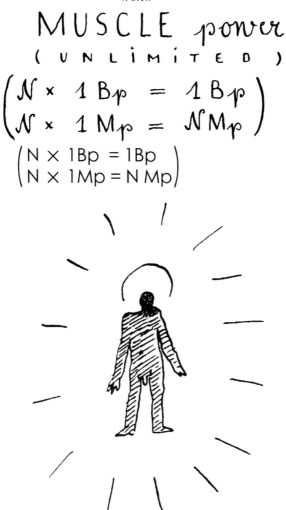

HOMO FABER
人体

两个世界

MIMICRY
creative or slavish
模仿
创造性的 或 无创造性的

	model (message) 模式 (主题)	
I love		I love
我爱		我爱

I love
I LOVE
I love
I LOVE
I love
I love
II LOVE
I LOVE

} IMITATION 模仿 {

I love
I lve
I lou
I liee
I l vi
I l ii

语义的模仿

SEMANTIC MIMICRY

sane
creative
sense producing

同样的
创造性
创作的感觉

图表的模仿

GRAPHIC MIMICRY

insane
slavish
non-sense producing

不同的
无创造性
创作的无感觉

LK 07

建筑和都市的空间

都市空间的四个基本类型

THE FOUR FUNDAMENTAL TYPES OF URBAN SPACE

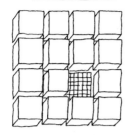

The Blocks are the result of a street and square pattern...
街区是由道路和用地的模式形成的

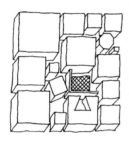

The streets and squares are the result of the position of the blocks...
街道和用地是由街区的状况决定的

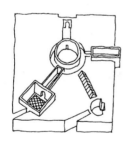

The streets and squares are precise formal types, the block is a result
街道和用地是精确的形式类型，而街区是一种结果

The objects do not form a describable space,
仅有建筑物不能形成明确的空间，
the public space is an accidental left-over
公共的空间是偶然剩余的

LK 77

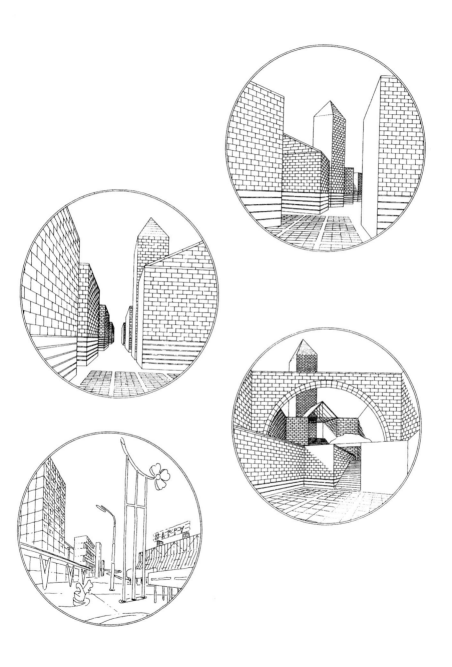

建筑和都市的空间

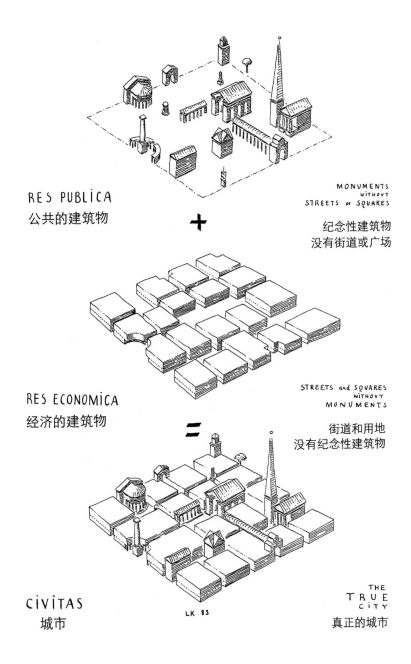

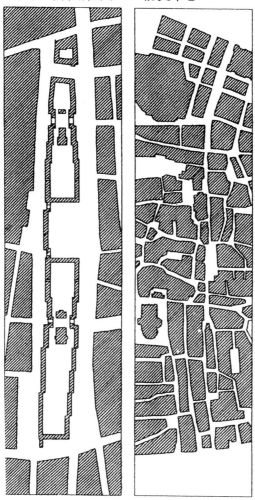

VIENNA
KARL-MARX HOF   HISTORICAL CENTER
维也纳
卡尔-马科斯霍夫   历史中心

3个建筑师　3 ARCHITECTS　　345 ARCHITECTS　345个建筑师
3公顷　　　~ 3 HA ~　　　　~ 3 HA ~　　　3公顷
　　　5.1 KM · PUBLIC FRONTAGE　18.5 KM
　　　　5.1公里 · 公共的空地 · 18.5公里

建筑和都市的空间　133

RELATIVE QUANTITY
公共空间的相对数量

1850 ~ 1920
1850～1920年

1945 ~ 1968
1945～1968年

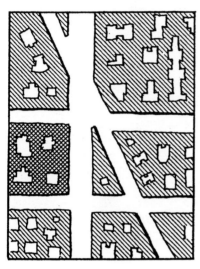 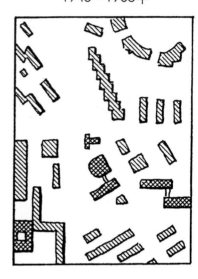

15 ~ 20 %
15%～20%
TOO LITTLE PUBLIC SPACE
太小的公共空间

70 ~ 80 %
70%～80%
TOO MUCH PUBLIC SPACE
太大的公共空间

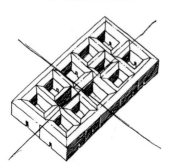 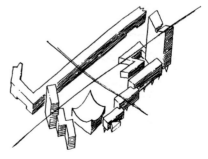

15 - 20% of public space
15%～20% 的公共空间

70 - 80% of public space
70%～80%的公共空间

公共空间的相对数量
# of PUBLIC SPACE

1970 ~ 1980 1970~1980年　**OPTIMUM** 最合适的

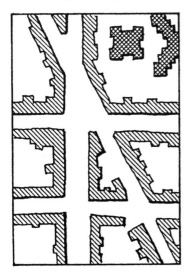
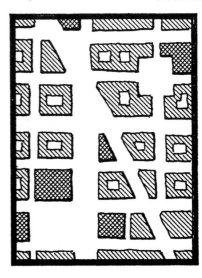

50 ~ 60 %
50%~60%

25 ~ 35 %
25%~35%

TOO MUCH SEMI-PUBLIC
过多的半公共空间

THE GOOD PROPORTION
好的比例

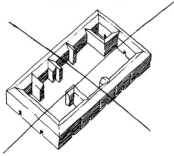
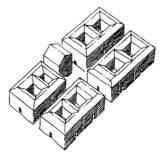

50-60% of public space
50%~60%的公共空间

25-35% of public space
25%~35%的公共空间

建筑和都市的空间

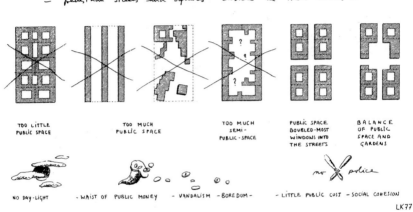

不仅柏林的街区是错误的，尺度也是错误的
不仅临街的公寓是错误的，庭院里的也是错误的
不仅街道的长度是个问题，街区的长度也是个问题
汽车交通停留在现有的街道里，
人行道和用地限制了新的街区

| 太小的公共空间 | 太大的公共空间 | 过多的半公共空间 | 公共空间大部分的窗子开向街道 | 公共空间与花园的均衡 |

没有日光　公共财物的浪费　肆意破坏　无聊乏味　很小的公共成本　社会凝聚力

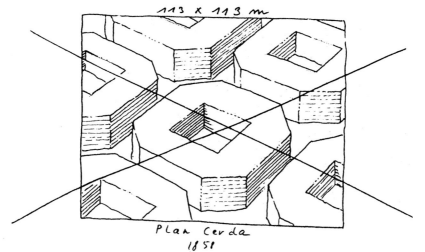

Cerda 的规划
1851年

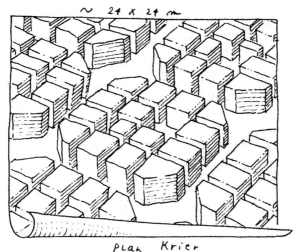

克里尔的规划
1976年

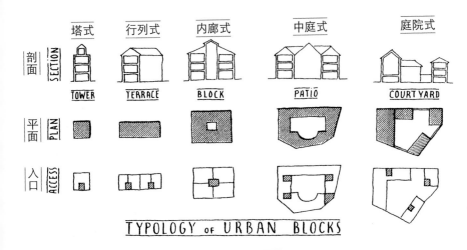

TYPOLOGY of URBAN BLOCKS

都市街区的类型

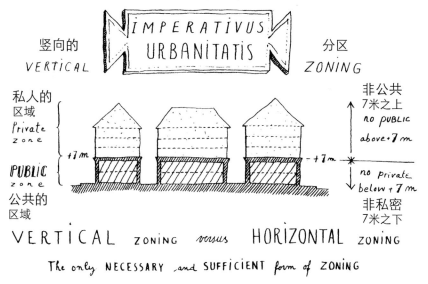

竖向分区　VS　水平分区
分区的唯一必需与必然的形式

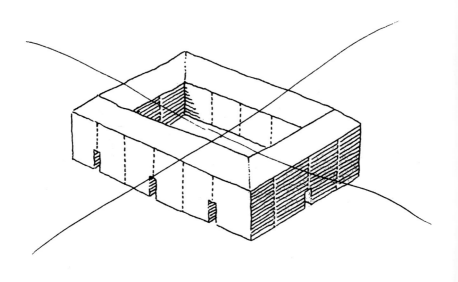
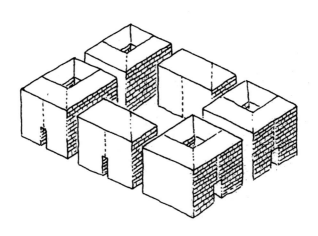

立面对称

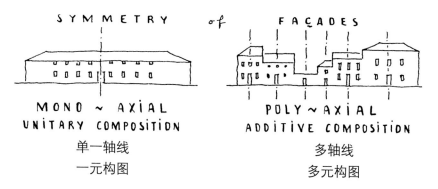

单一轴线
一元构图

多轴线
多元构图

小镇景观的活力

ANIMATED FACADES
POOR PAVING

富有活力的立面
简陋的铺装

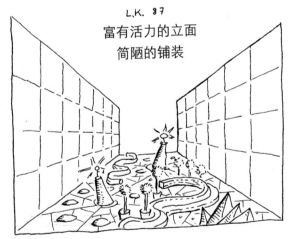

POOR FACADES
AGITATED PAVING

简陋的立面
纷乱的铺装

近似的必要
NECESSITY OF PROXIMITY

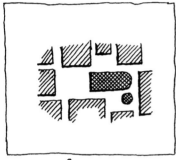

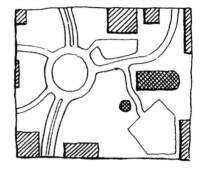

公共空间
作为市政区域

公共空间
作为交通区域

住区中社宅的布置

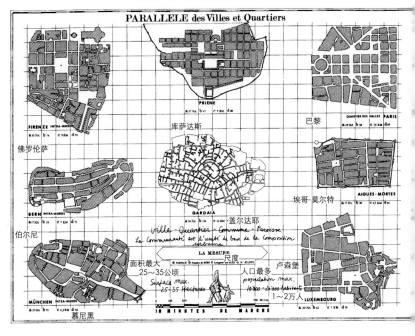

欧洲的城市

(前工业--)

一个城市根据其大小来确定

<u>社区的大小</u>

(宗教-军事-商业)

美国的城市

(北部)

前综合体

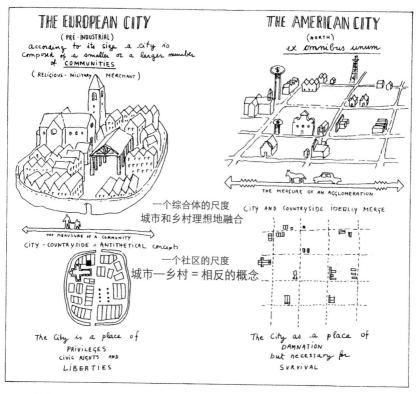

一个综合体的尺度
城市和乡村理想地融合

一个社区的尺度
城市—乡村＝相反的概念

城市是一个特权之地，市民
的权利与自由

城市是一个诅咒之地但是
对生存又是必需的

伦敦城

A alternative future
A 可选的未来

伦敦城

B

可持续的和不可持续的建筑高度

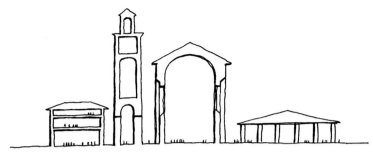

Low buildings & High ceilings
低矮的建筑物 & 高大的顶棚

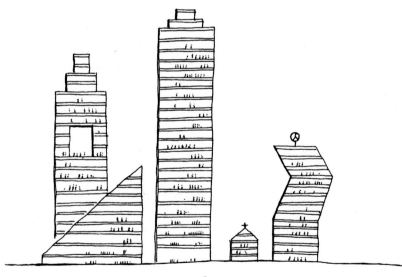

High buildings & Low ceilings
高大的建筑物 & 低矮的顶棚

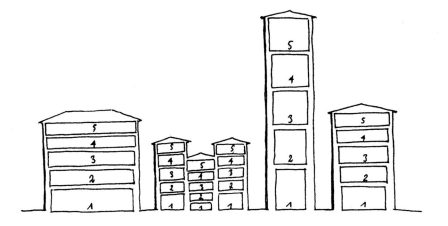

Limited Number of Floors
No height limit ~ Maximum Variation of Building & Ceiling height

VARIED SKYLINE

限定楼层的数量
没有高度限制～建筑物&顶棚高度
的最大可能的变化的

多样的轮廓线

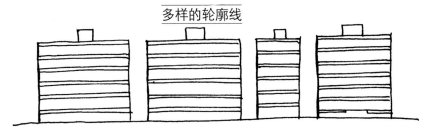

Limited Building Height
Maximum realisation of rentable floors - Minimum Ceiling Height

UNIFORM SKYLINE

限定建筑物的高度
最大限度地获得可使用的
楼层～ 最小的顶棚高度

单一的轮廓线

可持续的和不可持续的建筑高度

① Land-scraper　② Sprawler　③ Sky-scraper
①景观　　　　②房子　　　　③摩天楼

3 Basic Modern Building-types
现代建筑的三种基本建造类型

LK 87

3 Forms of Environmental Pollution
FUNCTIONAL MONOTONY >>> ARCHITECTURAL PATHOLOGIES
三种污染环境的形式
功能单调 >>>> 建筑病理

MONOFUNCTIONAL ZONING ⇒ MEGASTRUCTURES
单一功能分区 ⇒ 超级建筑

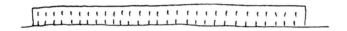

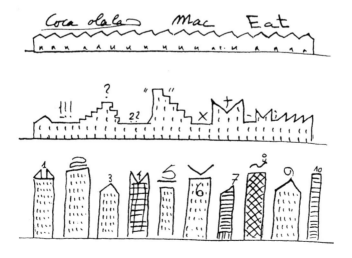

可持续的和不可持续的建筑高度

彼得·潘式的情景
PETER PAN SCENARIO

综合症

till when?   到何时?

? BUILT OUT PEAK ?   ? 打造巅峰 ?

1920年 1920

2020年 2020

2120年 2120

2220年 2220

OIL PEAK — OPTIMISTIC FORECAST — 2220

then what?

石油峰值 — 最优化的森林 — 2220年

然后又是什么呢?

过高容积率的长期效应
Long Term effects of excessive Plot-Ratio ordinances
Manhattan Logic
曼哈顿逻辑

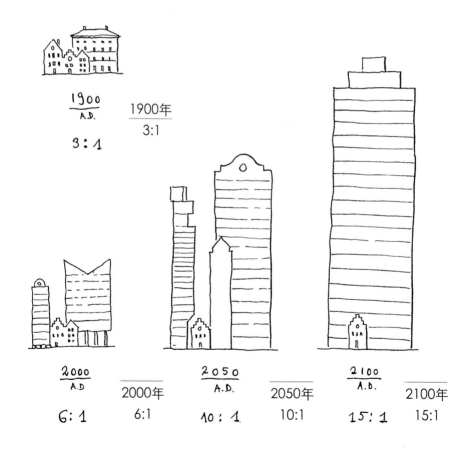

可持续的和不可持续的建筑高度

THE FATAL LOVE~STORY
致命的爱情故事

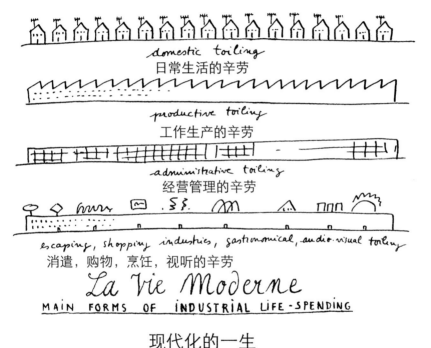

可持续的和不可持续的建筑高度

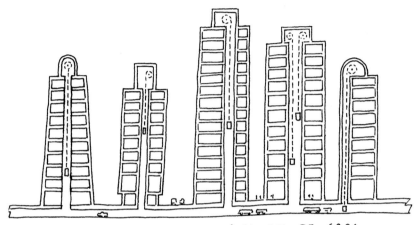

SKYSCRAPERS as VERTICAL CUL-DE-SACS or NETWORK CONGESTORS

摩天楼—竖向的尽端式道路

网络堵塞

两种主要的网络堵塞
（网络寄生）

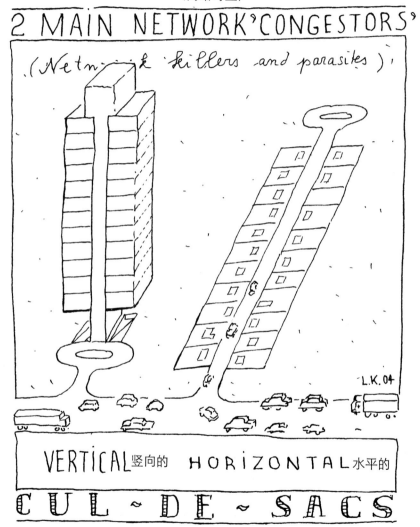

可持续的和不可持续的建筑高度

建筑的
勃 起
（已知无药可治）

architectural
PRIAPISM
(no cure known)

PRIAPUS HUBRIS   PRIAPUS ~ NEMESIS
                        LK · 2004

勃起　傲慢自大   勃起 ～ 上天的惩罚

THE TOWER~DRIVE
controled ~ uncontroled
高层建筑的压力
控制的～无控制的

TOWER-LESS SPRAWL
FUTURELESS HORIZONTALITY

"towerless"
"无高层的"

无高层建筑的扩展
没有前途的水平发展

TRUE SKYSCRAPER CITY
HORIZONTAL CITY WITH VERTICAL ACCENTS
LK 06

"tower-ful"
"有高层的"

真正的摩天楼城市
有竖向重点的水平城市

SPRAWLING TOWERS
FUTURELESS VERTICALITY

"tower-sick"
"病态的高层"

蔓延的高层建筑
没有希望的竖向发展

转换的建筑与都市肌理

HOW MUCH CLASSICAL & Vernacular IS NEEDED TO MAKE THE GOOD CITY?

要造出好的城市需要多少古典建筑和乡土建筑？

100% 古典的
100% CLASSICAL

C

好的城市
THE GOOD CITY

V

100% 乡土的
100% Vernacular

美术　乌托邦
芝加哥　白色城市 1893年
凡尔赛　城堡+公园
拿破仑三世—维多利亚式
帝国
罗马　战神广场
北京　紫禁城
雅典　卫城

文艺复兴的理想城市

BEAUX-ARTS UTOPIA

CHICAGO WHITE CITY 1893

VERSAILLES CASTLE + PARK
NAPOLEON III - VICTORIANA

IMPERIALISME
ROME CAMPO MARTIO
FORBIDDEN CITY BEJING
ACROPOLIS

RENAISSANCE IDEAL CITIES

VENICE
ROTHENBURG O.T.
CESKY - KRUMLOV
ATHENS CLASSICAL AGE
WILLIAMSBURG

TRADITIONAL VILLAGES

PRIMORDIAL HAMLETS
LASCAUX
ANIMAL ARCHITECTURE

威尼斯
罗腾堡
捷克 - 克鲁姆
威廉斯堡

传统村庄

拉斯科

动物的建筑学

TUNING OF URBAN NETWORK & ARCHITECTURE
都市网络 & 建筑的转换

© LK 03

- 都市 URBAN. [U] □ URBANISM 都市主义
- 建筑 ARCH. [A] ☒ ARCHITECTURE 建筑学
- 本地 VERN. [V] ☒ VERNACULAR 乡土的
- 等级 CLASS. [C] □ CLASSICAL 古典的

- NETWORK OF PUBLIC ROUTES & SPACES
  公共路线与空间的网络
- PUBLIC / CIVIC — PRIVATE DOMESTIC
  公共的/市政的 —— 私家的
- INFORMAL – ORGANIC GEOMETRY — PROSE
  非正式的 —— 有机几何 散文
- FORMAL – EUCLIDEAN GEOMETRY — POETRY
  正式的 —— 欧氏几何 诗

			Examples	中文
I	URB. VERN. 都市.乡土 ARCH. VERN. 建筑.乡土	UV AV	• VILLAGES • HAMLETS • PUEBLOS • FARMS • "ENSEMBLES SPONTANÉS" • PORT-GRIMAUD	乡村・原始村落・印第安人的村子 形式・"自发的整体性"・Grimaud港
II	URB. VERN. 都市.乡土 ARCH. CLASS. 建筑.等级	UV AC	• ATHENS ACROPOLIS • AGORA • REPUBLICAN FORUM ROME • OLYMPIA • NARA TEMPLES • MONTI SACRI	雅典卫城・市场 罗马共和国广场 奥林匹亚・奈良神社 圣山
III	URB. VERN. 都市.乡土 ARCH. VERN. + CLASS. 建筑.乡土+等级	UV AV+C	• VENEZIA • LUCCA • UZES • SAN GIMIGNANO • ÎLE DE LA CITÉ • ROTHENBURG • DINKELSBÜHL • SIENA • ASSOS • BELLAGIO • SEVILLE	威尼斯・卢卡 圣基米尼亚诺・城市 罗滕堡・丁克尔斯比尔・锡耶纳 阿索斯・拜拉吉奥・塞维利亚
IV	URB. CLASS. 都市.等级 ARCH. VERN. 建筑.乡土	UC AV	• BARRACKS • CONCENTRATION CAMPS • PRISONS • "VILLE RADIEUSE" • HILBERSEIMER BERLIN • INDUSTRIAL UTOPIAS • SPANISH ENSANCHES • NEW DEAL HOUSING	兵营・集中营 监狱・"拉第乌斯镇" 希尔伯斯玛的柏林・工业的乌托邦 西班牙的扩展・新的分配住宅
V	URB. CLASS. 都市.等级 ARCH. CLASS. 建筑.等级	UC AC	• IMPERIAL CAPITALS • BATH • VERSAILLES • BABYLON • KARNAK • FORBIDDEN CITY • RENAISSANCE UTOPIAS • BEAUX-ARTS • TURIN • HAUSSMAN-PARIS • SPEER-GERMANIA	帝国首都巴斯・凡尔赛 巴比伦・紫禁城・卡纳克 文艺复兴的乌托邦・美术都灵 奥斯曼・巴黎・施佩尔日耳曼
VI	URB. CLASS. 都市.等级 ARCH. VERN. + CLASS. 建筑.乡土+等级	UC AV+C	• PRIENE • TIMGAD • KYOTO • WINDSOR FL. • CHARLOTTESVILLE • WILLIAMSBURG • YALE CAMPUS • JEFFERSON GRID • SALINES DE CHAUX • LAW OF INDIES • BASTIDES TOWNS • FIRENZVOLA	普里埃尼・提姆加德・东京温莎・佛罗里达 夏洛特镇・威廉斯堡・耶鲁校园・杰斐逊网 皇家制盐所・随处可见的小镇 斐伦兹沃拉
VII	URB. VERN. + CLASS 都市.乡土+等级 ARCH. VERN. 建筑.乡土	UV+C AV	• HAMPSTEAD GARDEN SUBURB • SUBURBIA • MODERNIST NEW TOWNS • UK • FRANCE • USSR • CHINA • BRASILIA	汉普斯特德花园郊区 郊区・现代派的新城 英国・法国・俄罗斯 中国・巴西
VIII	URB. VERN. + CLASS 都市.乡土+等级 ARCH. CLASS. 建筑.等级	UV+C AC	• PALMYRA • LEPTIS • CHANTILLY CASTLE • RESIDENZ STÄDTE SCHWETZINGEN • POTTDAM • VATICAN-CITY • CESKY-KRUMLOV • BORDEAUX • KREMLIN	巴尔米拉・莱普提斯・尚蒂伊城堡・居住城市施威琴根 波茨坦・梵蒂冈城・克鲁姆洛夫・波尔多・克里姆林宫
IX	URB. VERN. + CLASS. 都市.乡土+等级 ARCH. VERN. + CLASS. 建筑.乡土+等级	UV+C AV+C	• PIENZA • NOLLI-ROME • PRAGUE • ISE • ISTANBUL • CAIRO • PRAGUE • CUSCO • SAMARKAND • KATMANDU • LHASSA • NEW URBANISM • POUNDBURY	皮恩扎・诺里・罗马・布拉格 伊斯坦布尔・开罗・库斯科 德累斯顿・库斯科・撒马尔罕 加德满都・拉萨 新都市主义・庞德伯里

转换的建筑与都市肌理

都市建筑的转换
《你的》
都市主义

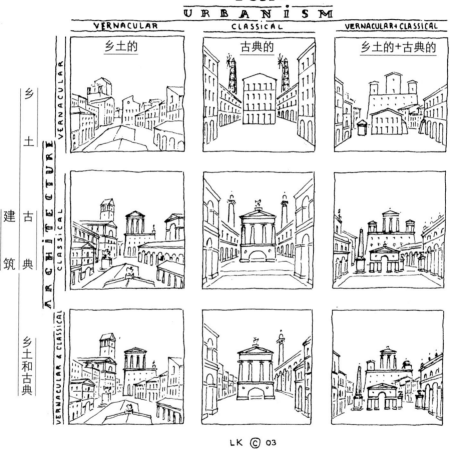

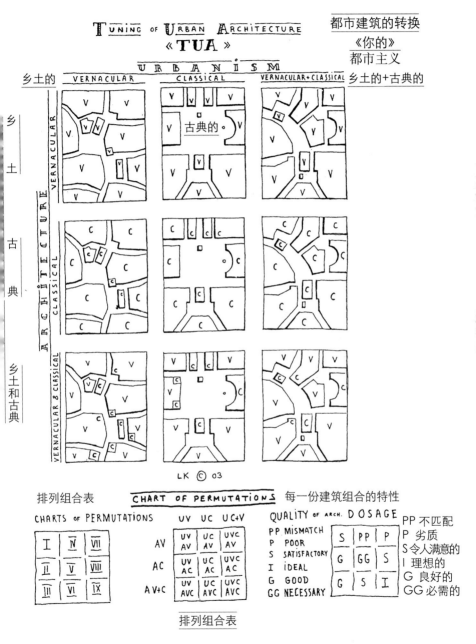

转换的建筑与都市肌理

TO MAKE a CITY
MIXED USE
IS NECESSARY BUT NOT SUFFICIENT
a condition

制造一个城市
用途的混合
是必需的但不是充分的
情况

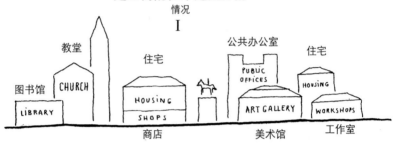

CITY as LAND ~ FLOTILLA
x number of architects

城市—陆地~舰队
X位建筑师

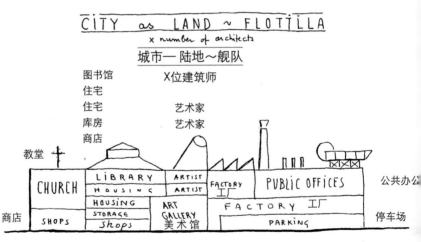

"CITY" as LAND ~ LINER
1 architect

"城市"—陆地~邮轮
1位建筑师

TO MAKE a CITY
MIXED USE
IS a NECESSARY BUT NOT a SUFFICIENT condition
II

制造一个城市
用途的混合
是必需的但不是充分的情况
II

图书馆　庙宇　住宅 商店　公共办公室　房子　市场

LIBRARY | TEMPLE | HOUSING SHOPS | PUBLIC OFFICES | HOUSES | MARKET

TYPOLOGICAL ORDER
FUNCTIONAL ≡ ARCHITECTURAL VARIETY

类型的秩序
功能的 = 建筑的多样

图书馆　庙宇　住宅 商店　公共办公室　房子　办公室 市场

LIBRARY | TEMPLE | HOUSING SHOPS | PUBLIC OFFICES | HOUSES | OFFICES MARKET

BUROCRATIC ORDER
FUNCTIONAL VARIETY ≫≪ ARCHITECTURAL UNIFORMITY

官僚主义的秩序
功能的多样 ≫≪ 建筑的单一

住房　办公室 公共办公室　房子　公寓大楼 办公室 市场

CONDO | 公寓大楼 | 住宅 商店 HOUSING SHOPS | P-HOUSE OFFICES PUBLIC OFFICES | HOUSES | CONDO OFFICES MARKET

图书馆 LIBRARY | 庙宇 TEMPLE

OVER DEVELOPMENT - MANHATTANISM
FUNCTIONAL VARIETY ≫≪ ARCHITECTURAL VARIETY

过度发展 - 曼哈顿主义

功能的多样 ≫≪ 建筑的多样

转换的建筑与都市肌理　169

都市构图的建筑 转换
ARCHITECTURAL TUNING OF URBAN COMPOSITION
⟨ vernacular & classical ⟩
<乡土的 & 古典的>

vernacularissimus
AUSTERITY VERNACULAR
乡土

乡土的朴素

vernacular & classical
CULTURAL APOGEE
乡土 & 古典

文化的制高点

CLASSICISSIMUS
IMPERIAL CARNIVAL CLASSICISM

L.K. 06
古典

帝国嘉年华式的古典主义

APPLYING & SIZING OF CLASSICAL and vernacular MODES

古典的应用与排列和乡土的模式

well-applied & well-sized
良好的应用 & 良好的尺寸组合

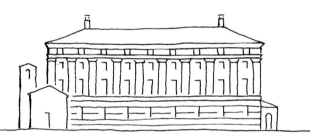

mis-applied & mis-sized
不好的应用 & 不好的尺寸组合

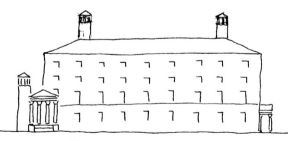

well-applied & mis-sized
良好的应用 & 不好的尺寸组合

LK 05

PRIVATE IMPERIALISM

个体的帝国主义

GOOD PRIVATE/PUBLIC AFFAIRS

良好的个体/公共状态

PUBLIC IMPERIALISM

公共的帝国主义

LK 03

我是一个生活的好地方

architectural SPEECH

建筑的演讲

我…我…我…我…我…是…是…是…是…是…是…

architectural STUTTER

建筑的口吃

LK 99

形式与同一

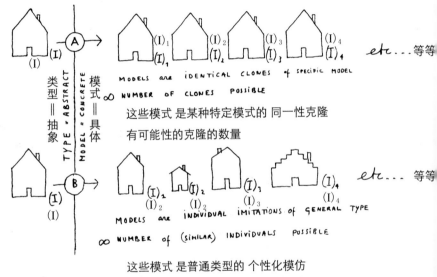

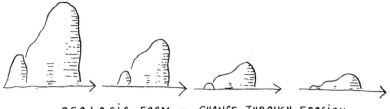

GEOLOGIC FORM ~ CHANGE THROUGH EROSION
"nature"

地理形式 ～ 因侵蚀而改变
"成熟"

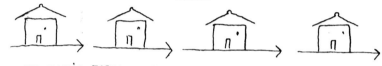

TECTONIC FORM ~ PERMANENCE THROUGH ARCHITECTURE
EROSION MASTERED BY TECHNICAL ARTIFICE
ANALOGICAL IMITATION
of nature

技术的形式 ～ 因建筑而持久
侵蚀 由技术手段控制
成熟的类比模仿

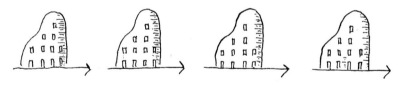

GEOLOGIC FORM SIMULACRUM ~ FROZEN EROSION
EROSION CHECKED IN ERODED FORM
MECHANICAL IMITATION
of nature

地理形式的拟像 ～ 凝固的侵蚀
以错误的形式控制的侵蚀
成熟的机械模仿

房间 或 走廊 的类型
结构的细胞 或 动脉
不能再混为一谈了！！！！拜托

TYPOLOGIES of ROOMS or CORRIDORS
of cellular or arterial structures
not to be confused EVER AGAIN !!!!! please

DOMINANT VOLUME & ROOM CORRESPOND
perceivable at one glance

主体的体量 和 相应的房间
一目了然的

BUILDINGS AS LABYRINTHINE
circulation structures

迷宫般的建筑物
循环的结构

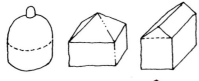
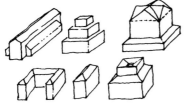
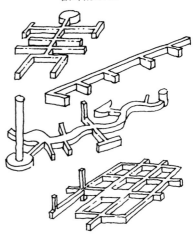

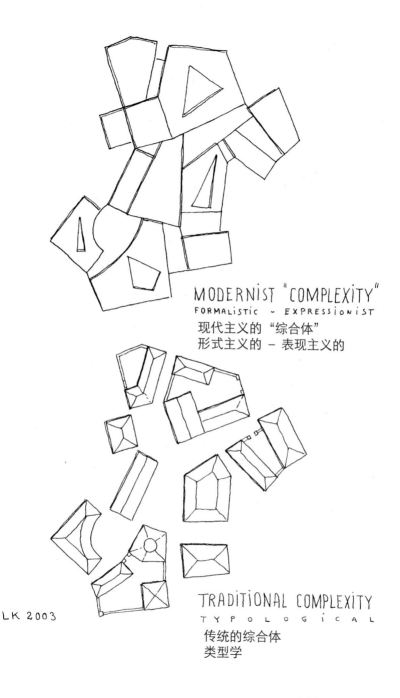

形式与同一

同一
UNIFORM

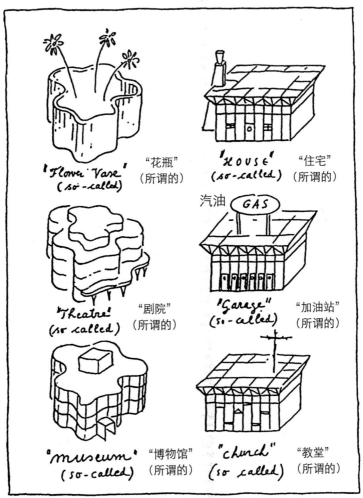

SO~CALLED "OBJECTS"
所谓的 "单体建筑"

对 *versus* 形式 # FORM

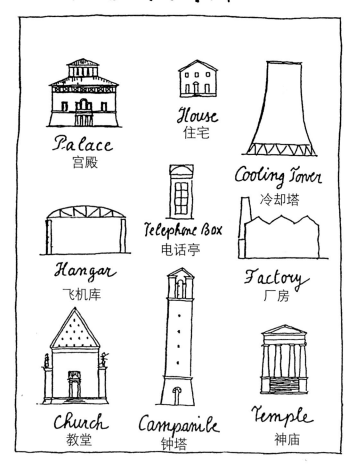

NAMEABLE OBJECTS
名副其实的 单体建筑

家用的 & 公共的　规模　质量的 或 数量的

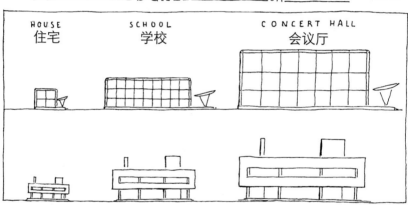

不要忘了你的 重力 限制
DON'T FORGET YOUR GRAVITY CHECK

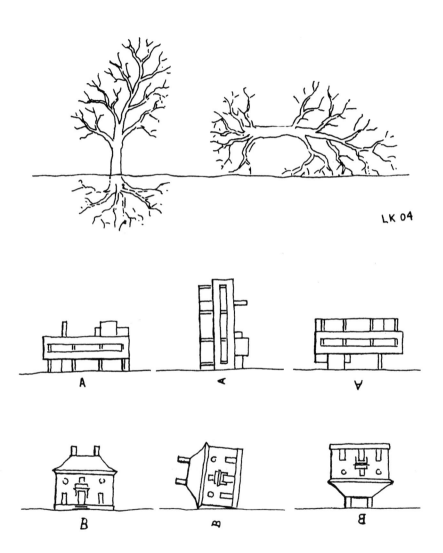

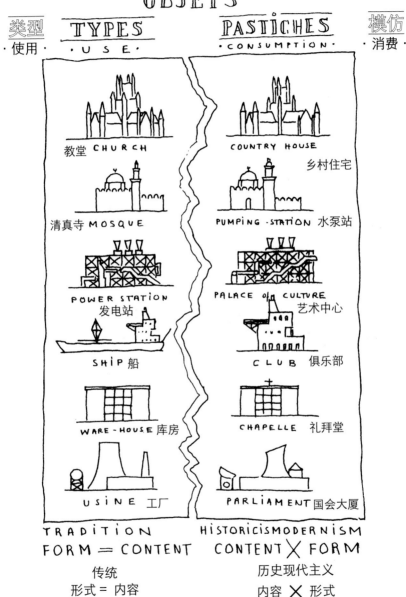

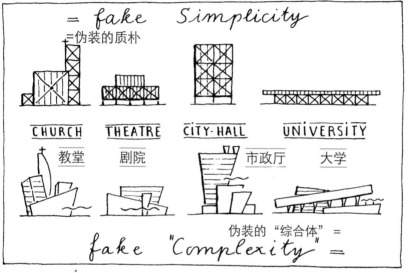

形式与同一

这个学校坚决不能做成只有<u>一个</u>出入口的<u>单体建筑</u>。

This school must not be composed like a <u>single</u> building with one entrance

TWO KINDS OF MONSTERS
两种庞然大物

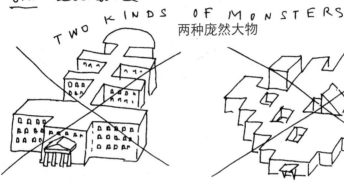

AUTHORITARIAN LOOK　专制的 看起来

DEMOCRATIC LOOK　民主的 看起来

This school will be composed like <u>a city</u> with small and big buildings according to their importance

这个学校将做成一个由重要性不同的大大小小的建筑物组成的<u>城市</u>

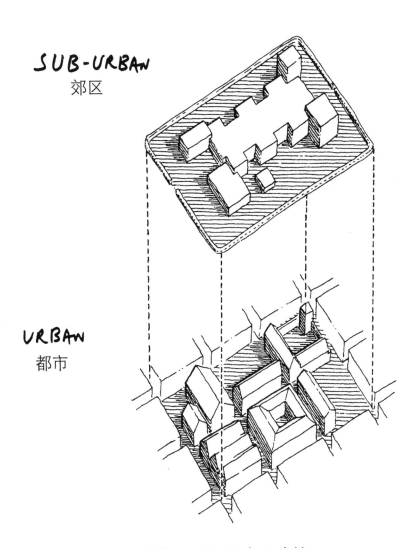

学校 + 城市的中心建筑

SCHOOL + civic center buildings

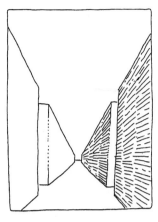

UN-FOCUSED　　　　FOCUSED
无焦点的　　　　　有焦点的

STREET
街道

As CORRIDOR　　　As PLACE
作为通道　　　　　作为场所

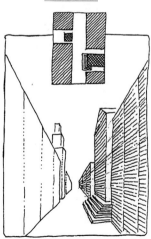 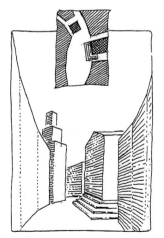

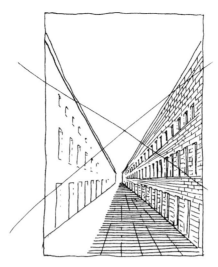

The Prison
Two Streets
监狱
两条街道

The City
城市

形式与同一

保存与维护

城市的重构

手 工 艺 的 城 市

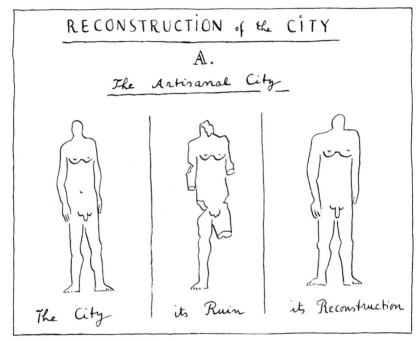

城市　　　　　　废墟　　　　　　重构

城市的重构
Ⓑ
工 业 的 反 城 市

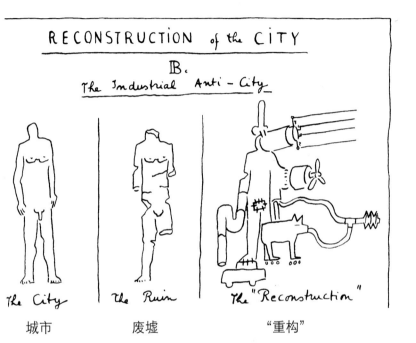

城市 废墟 "重构"

保存与维护

UNACCEPTABLE LIMITATIONS ON THE CAPITALS SKYLINE

不被认可的限制
关于
首都的天际线

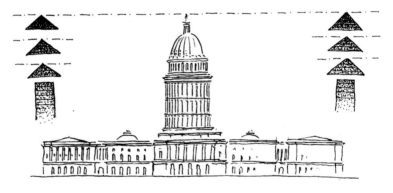

THE "UPWARD SOLUTION"
NEGOTIABLE

LK 04

向上的"解决方案"
可协商的

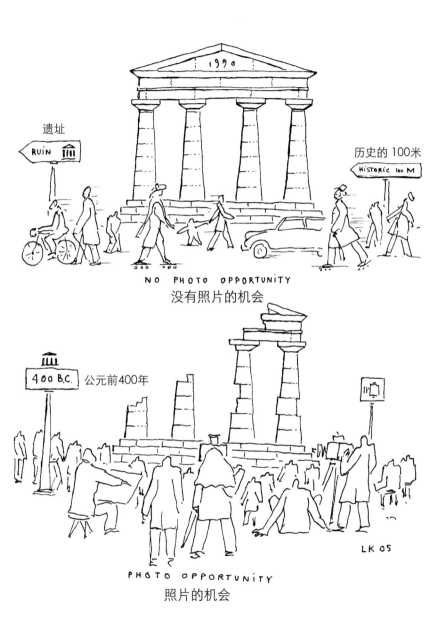

保存与维护 195

LK 85

Ex ~ tensions
过分的张力

复原 保守的 对 创新的
Conservative versus Creative Restauration

Derelict Modernist Masterpiece
废弃的现代主义作品

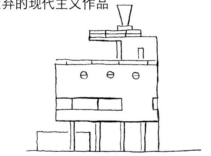

Conservative Restauration (Nostalgic)
保守的复原（怀旧的）

Creative (forward looking) Restauration
创新的复原（向前看的）

保存与维护

通过对比来尊重
Respect through Contrast

Reichstag

德国议会大厦

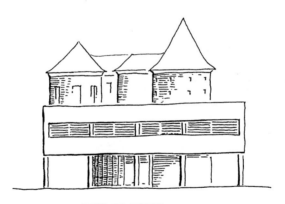

Villa Savoye

萨伏伊别墅

"Charter of Venice" and "Conservation has gone too far" principles ecumenically applied

《威尼斯宪章》和"保护已经早已消失的"原则普遍地应用

LK 99

Charta of Venice applied to the restoration of Gozzoli - fresco LK 1986
《威尼斯宪章》适用于戈佐利壁画的修复

保存与维护 199

猜一猜谁正在未来的政府中
给予你关于伦敦的发言权

G.L.C. PROPAGANDA UNIT 1985
大伦敦市议会传单 1985年

猜一猜在过去的政府里谁不曾给你发言权

GUESS WHO OFFERED YOU NO SAY IN THE PAST GOVERNMENT FOR LONDON.

This is what the GLC did

to the Greater Part of London.

Londoners look at your city!!

这是大伦敦市议会为大部分的伦敦所做的事情。

伦敦人看看你们的城市！！

L.K. OPINION 1985

L·克里尔的观点 1985年

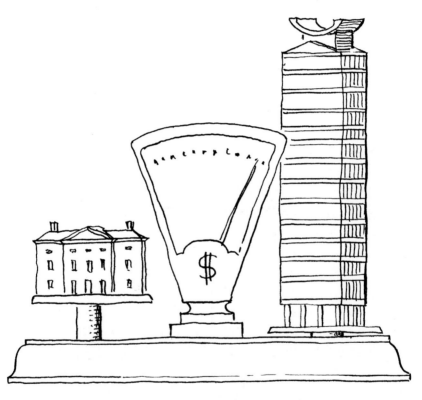

Conservation versus Over-Development

保护 VS 过度发展

PRESERVED HISTORIC CITY
保留下来的历史城市

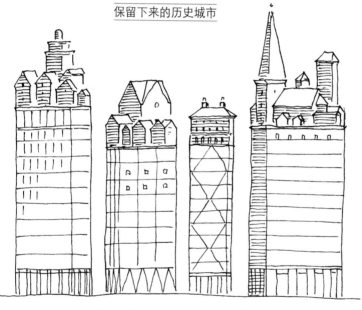

HISTORIC PRESERVATION VIA LIFT~OFF

LK 2005

历史的保留
通过
抬升

保存与维护

美丽的部件
BEAUTIFUL PARTS

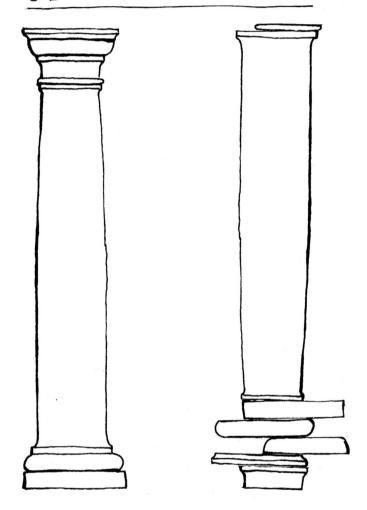

STABLE × ASSEMBLAGE × UNSTABLE
SUSTAINABLE   UNSUSTAINABLE

可持续的   稳定 × 装配 × 不稳定   不可持续的

保持 & 文化
MAINTENANCE & CULTURE

~ 1870   ~ 1984

from CLASSICAL to vernacular
observed in Belsize Park - London

从古典到乡土
在伦敦Belsize公园观察所得

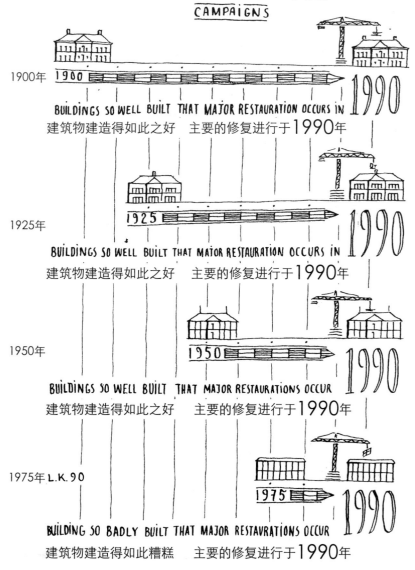

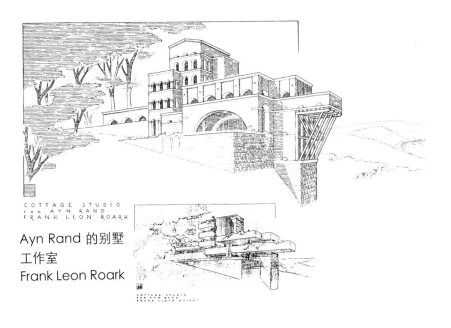

Ayn Rand 的别墅工作室
Frank Leon Roark

保存与维护

建筑病理

住宅的谱系
Genealogy of the House

30 A.D.
公元30年

1030 A.D.
公元1030年

1830 A.D.
公元1830年

1930 A.D.
公元1930年

2030 A.D.
公元2030年

TEMPORARY REFUSAL of the ARCHETYPE

LK 88

对原型的暂时拒绝

HOUSE
住宅

HOUSES
很多住宅
LK 82

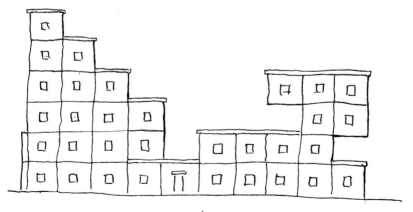

HOUSING
公共住房

建筑病理

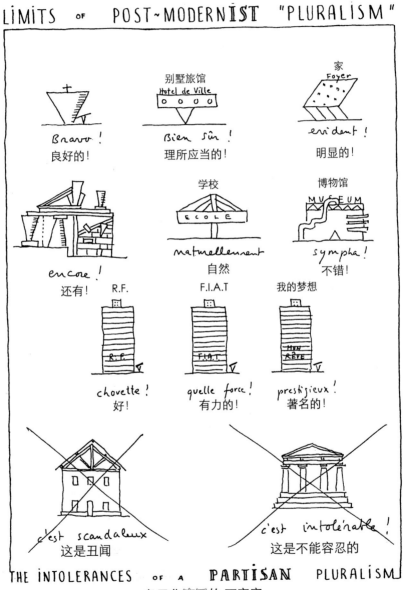

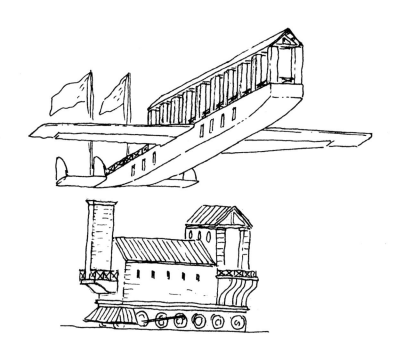

INNOVATION = CONFUSION OF GENRE

Let us suppose that one day an innovation fever was to befall engineers and would dictate them to abandon the means of their profession; would it then, I ask you, be less idiotic to ask the traveller to design his own aeroplane than it is to ask the inhabitant to invent the form of his house?

创新 = 类型的混乱

让我们假设有一天创新的狂热会与工程师们遭遇并将使他们放弃他们的专业手段。我问你，去要求乘客设计他们的飞机不比要求居民去发明他们的住房的形式更愚蠢吗？

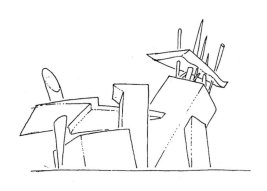

Award-winning Villa before Earthquake

备受赞誉的别墅建筑
地震之前

Award-winning Villa after Earthquake

备受赞誉的别墅建筑
地震之后

LK 89

轮廓线设计
新的水平线.com

CUTTING HEDGE DESIGN
the new horizon____.COM

"THE AGGRESSSOR"
"侵犯"

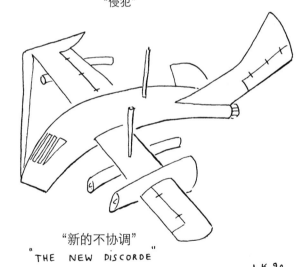

"新的不协调"
"THE NEW DISCORDE"

LK 90

why should cars and planes forever remain classical, symetrical etc........?

为什么汽车和飞机永远保持着古典、对称、……？

建筑病理　215

我们的 办公室景观
unsere Bürolandschaft

WHAT IF N° 10 DOWNING STREET HAD NOT BEEN REBUILT BY RAYMOND ERITH? LK 85

如果唐宁街10号没有被Raymond Erith重建会是什么样?

建筑病理

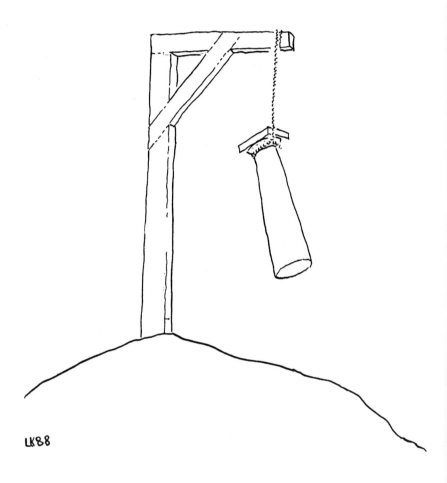

The Nüremberg-Tribunal of Architecture
建筑的纽伦堡法庭

mockba
(1933)

莫斯科
(1933年)

Totalitarian
male & female
sex ~ symbols

极权主义

男性 & 女性
性别 ~ 象征

Berlin
(1939)

柏林
(1939年)

建筑病理

艺术 & 工业技术
ART & INDUSTRIAL TECHNIQUE

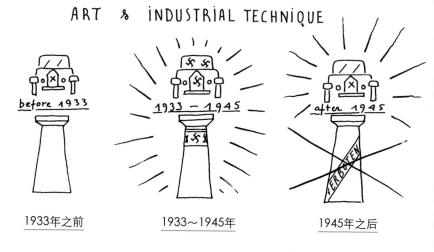

1933年之前　　　1933~1945年　　　1945年之后

INDUSTRIAL TECHNIQUE HAS WON THE WAR
BECAUSE ART IS A DECEITFUL
WEAPON　　　　　　　　　　　　LK81

oh GERMANY * BEWARE OF ART AND COLUMN

because only "TECHNOLOGY" will render invincible
* and others

工业技术已经赢得了战争
因为艺术是一种机巧的
武器

噢 德国 小心 艺术和柱子
因为只有"技术"将不可改变地给予

Speer à Rio
施佩尔 在里约

Speer à Berlin
施佩尔 在柏林

Speer à New York
施佩尔 在纽约

建筑病理

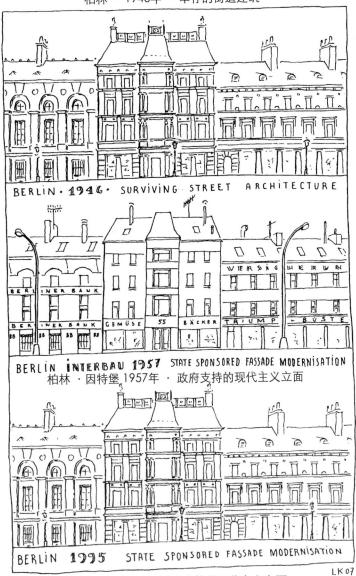

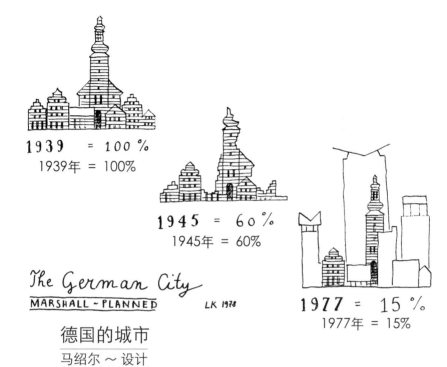

建筑病理 223

世界观
WORLD VIEWS

MODERNIST　　　　　　　　CLASSICAL

现代派　　　　　　　　　　古典的

精神王国隐藏的必然结果
THE HIDDEN COROLLARY of mental empires

"我是我的灵感的唯一来源……"。"我是唯一受到我自己影响的人……"。"上帝已经把我发送到……"。"我和其他任何人都不相同……"。"我只受事件的影响而不受任何人的影响,……"。"我是第一个……"。"我是你的……"

"I am my only source of inspiration"…"I am only influenced by myself"…"God has sent me to.."
"I am unlike any other"…"I am not influenced by anyone, only by events"…"I am the first"…"I am your salv.."
"The

"Those who are not for me are against me".
"I am the beginning and the End of… Philosophy, science, architecture, mathematics, cooking, grammer"
LK 06

"那些不支持我的人就是反对我。" "我是…哲学的创始者和终结者,因此,建筑、数学、烹饪、语法……"

建筑病理 225

现代主义的文化

la culture modern(ist)e

《 MARQUE DEPOSÉE 》

depuis 1920

LE COURAGE

《品牌的构成》

1920年

勇气

LA NOSTALGIE

怀旧

LK 99

可笑的"平屋顶"故事

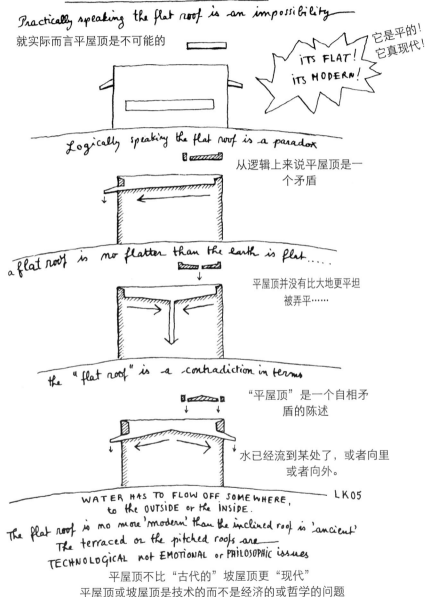

建筑病理

你，先生，已经超出了倾斜度限制…
…以法律的名义，我必须…………
严肃地处罚你……

A SUFFERING MINORITY GROUP VENTING A LONG STANDING GRIEVANCE IN ORDERLY MANNER

少数团体以有秩序的方式发泄忍耐已久的不满

建筑病理

世界的顶棚

WORLD CEILINGSCAPE

文化的 CULTURE-SCAPE

PRIVAT-SCAPE 私人的

CHURCH-SCAPE 教堂的

OFFICE-SCAPE
办公室的

FACTORYSCAPE
工厂的

Dedicated to JAMES HILLMAN ~ LK 06
献给 James Hillman

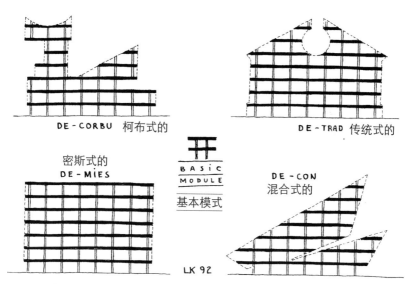

The Unified Field of New·Regional·Critical·Post·Structural·Anti·Classical POST & NEO~MODERNISM
(before installation of CURTAIN WALL)

新的·地区的批评的·后-结构的·反-古典的 同一领域
后现代主义 & 新现代主义
（在幕墙安装之前）

建筑病理